DISCIPLINES AND DOCTORATES

HIGHER EDUCATION DYNAMICS

VOLUME 16

SCOPE OF THE SERIES

Higher Education Dynamics is a bookseries intending to study adaptation processes and their outcomes in higher education at all relevant levels. In addition it wants to examine the way interactions between these levels affect adaptation processes.It aims at applying general social science concepts and theories as well as testing theories in the field of higher education research. It wants to do so in a manner that is of relevance to all those professionally involved in higher education, be it as ministers, policy-makers, politicians, institutional leaders or administrators, higher education researchers, members of the academic staff of universities and colleges, or students. It will include both mature and developing systems of higher education, covering public as well as private institutions.

Disciplines
and Doctorates

By

Sharon Parry
Southern Cross University, NSW, Australia

Springer

A C.I.P. Catalogue record for this book is available from the Library of Congress.

ISBN-10 1-4020-5311-8 (HB)
ISBN-13 978-1-4020-5311-5(HB)
ISBN-10 1-4020-5312-6 (e-book)
ISBN-13 978-1-4020-5312-2 (e-book)

Published by Springer.
P.O. Box 17. 3300 AA Dordrecht, The Netherlands.

www.springer.com

Printed on acid-free paper

This book is dedicated with love to Tony Becher,
who gave me a place in his tribe,
and taught me about the territory.

TABLE OF CONTENTS

FOREWORD

For many potential readers confronted with a new scholarly work, its context – its coordinates on the unfolding map of knowledge – is a substantial concern. With this said, the provenance of Sharon Parry's work is clear enough. In the broadest terms, it is unequivocally a contribution to the study of higher education. More precisely, it can be placed in the arena of microstudies (the field of enquiry which deals with the activities of individuals and small groups, as against macrostudies, which focus on higher education politics, economics, governance and the like). More specifically still, it belongs to the small but not insignificant body of research dealing with the similarities and differences between and within academic disciplines. Its proponents are liable to argue that it occupies the central core of research into higher education, providing an understanding of the inner workings of the academic world.

It is pertinent to ask what a work of this kind contributes to current understanding, or knowledge, or both – or how it fares in relation to the (obscurely connected) dichotomy between theory and practice. *Disciplines and Doctorates* comes out well in terms of all four requirements. Only a single example of each matching of the text against these criteria will be offered here, but the reader may note others as the argument proceeds.

Its contribution to the understanding of the doctoral process derives mainly from the strong emphasis on regarding it as a broadly cultural phenomenon. This gives rise to a series of apparently simple but previously neglected findings, one instance being the practice of supervisors to earmark promising students as entitled to "join the club." Recognising this helps to make sense of a number of hitherto puzzling practices.

A significant addition to existing knowledge rests in the substantial and wide ranging body of data drawn from the author's research interviews. In many cases key findings are given added credibility by the skilful use of multiple perspectives representing the differing views of students, supervisors and academic managers.

Although it is presented as a report on research findings rather than as a training manual, there are numerous pointers to what is good and what is bad practice in supervision, and what are rewarding or unrewarding strategies for doctoral students.

A significant development of current theory can be found in the author's innovative analysis of the linguistic patterns of doctoral theses in different groups of disciplines. The results of this analysis provide a fresh insight into the particular nature of each disciplinary group and draw out their connections to the cultural practices of those who subscribe to them.

It can fairly be said, in the light of these and other considerations, that this book is an important and original contribution to the literature on higher education. By way of an added bonus, the writing is illuminated by a pleasingly elegant style and an absence of obtrusive jargon.

What all this amounts to is a single exhortation: go ahead and read the pages that follow. You will not be disappointed if you do.

Tony Becher

ACKNOWLEDGEMENTS

This research was conducted with support from a grant from the Australian government's Evaluations and Investigations Program, and with support from Southern Cross University. A large number of participants – supervisors and doctoral candidates alike – made considerable contributions to the research, and their experiences, values and approaches were reported candidly and enthusiastically. Without those rich contributions to the database, it would not be possible to demystify the process of knowledge-making and knowledge reporting at the level of the research doctorate, which has for so long needed close examination. Many distinguished scholars of higher education policy have also made significant input to this work. For their advice and friendship and intellectual wisdom, I am very grateful. Finally, I acknowledge the long-term and considered contribution to this work of Professor Tony Becher, to whom this book is dedicated.

PREFACE

This book owes a large debt to two sources. The influence of Tony Becher on the theoretical foundations of this book is obvious. His insight and inspiration for the investigation it describes, and to the discussion it generates is substantial. Inspiration also derives from the very influential work of Maurice Kogan and Mary Henkel. These researchers have all contributed substantially to our knowledge of doctoral supervision and the nature of the learning that takes place at the level of the PhD. A second and vital contribution was made by the many doctoral supervisors and students who participated in the investigation. They gave generously their personal experiences and considered views. This book is intended to give voice to their knowledge and wisdom.

PART 1

KNOWLEDGE-MAKING IN DOCTORAL PROGRAMS

CHAPTER 1

INTRODUCTION

1. THE IMPORTANCE OF DOCTORAL STUDY

Doctoral programs have been proliferating in developed countries since the early 1990s. This conspicuous growth, accompanied as it has been by a remarkable increase in their diversity, is raising important questions concerning the standards represented by the doctoral award, the comparability of these standards across national and disciplinary boundaries, and the appropriateness of different forms of intellectual and practical support during doctoral candidature. While these questions need to be addressed, a more fundamental question, critical in importance as a basis for addressing all of the others, concerns the nature of the learning that, in fact, takes place during doctoral candidature. The question is: what are the forms of learning that are intrinsic to the experience of earning the academic title of "doctor"?

This book seeks to develop an understanding of what is involved in obtaining a doctorate. It addresses particularly the ways in which doctoral students acquire 'know-how' and confidence in a disciplinary area. Without this understanding, norms for quality in the design and implementation of doctoral programs are diffi- cult to establish, and reasons for the existence of so many different forms of doctoral experience are difficult to comprehend.

Much has already been written about the formal requirements and conventions of doctoral study: the nature of the doctoral investigation, acceptable thesis structures, appropriate reporting styles and acceptable progression schedules. There is also a growing literature on supervisory strategies and organisational models for support- ing doctoral supervision, some of which (Becher, Henkel and Kogan, 1994; Parry and Hayden, 1994; Parry, 1998; Parry and Hayden 1999; Delamont, Atkinson and Parry, 2000) has sought to interpret doctoral candidature as a period of socialisation to a specific disciplinary culture. Experiences of this socialisation across broad disciplinary areas have not, however, been examined at length and in depth in any of the literature to date. The purpose of this book is to fill in this part of the picture.

The research doctorate is a distinctive award. Earning it represents the attainment of a pinnacle of academic achievement. It holds more *caché* than nearly all other awards. It connotes mastery of a discipline area, confidence and agility in the mak- ing and reporting of new knowledge in a particular field, and 'know-how' in the construction of a sustained argument. A research doctorate conveys the acquisition of disciplinary savvy, as well as a confidence in using the knowledge and reporting tools of a discipline. These qualities, though often difficult to define and articulate, are nonetheless instantly recognisable by examiners of considerable standing in their fields.

1.1. Why look more closely at doctoral study?

Perhaps the primary reason for looking more closely at doctoral study is that the research doctorate remains one of the most desirable of all postgraduate awards. Its status is high, and it is universally recognised to provide an appropriate training for a research career. At the same time, concern is widely expressed about its often-narrow focus, and about long completion times and poor completion rates. These concerns have led to significant efforts to customise its form. Henkel (2000) reports that these efforts amount to the 'sciencification' of research training, a development that ignores the fundamental nature of knowledge production and reproduction in particular disciplinary areas – not to mention more than two decades of empirical research in this field.

There are other developments that make an examination of doctoral study timely. There is now more pressure on experienced researchers to supervise doctoral students, due simply to the increased numbers of students in doctoral programs. This pressure intensifies the search for more efficient ways of assisting the 'rite of passage' that doctoral candidature represents. This pursuit of economy in the research enterprise has inherent risks for the nature and quality of supervision, and these risks warrant a better understanding of how doctoral candidature is most effectively progressed.

Another development concerns the global trend towards increasingly specialised transdisciplinary and applied fields of research. New fields of research are cutting across long-held notions of disciplinarity and subject-area borders, as Gibbons and others (1994) have described. Doctoral students may well find themselves embarking on a research enterprise where the field of disciplinary specialisation (referred to hereafter as the specialism) is fairly novel, lacking in established traditions, perhaps with few indicators of what is acceptable. There may well be a need for the student to form controversial alliances among discrepant fields of knowledge to develop appropriate methods, techniques and theoretical frameworks for the study in question. In these circumstances, there is a need to better understand the learning process at the doctoral level so that systems, institutions and departments can effectively support doctoral research and give doctoral students clearer parameters concerning how they might to make good progress.

Yet another development concerns a recent proliferation in the kinds of doctoral awards available – among them coursework doctorates, professional doctorates, port-folio doctorates and the like. This proliferation is associated with the dispersal of research activity outside the confines of individual universities, involving closer relationships with industry, the community and the world of work. The task of locating doctoral study within a theoretical framework that informs its conduct, parameters and standards has never been more urgent. To this end, a useful distinction between the traditional research doctorate, the PhD or equivalent, and other forms of doctorate, particularly professional doctorates, is made by Bourner, Bowden and Laing (2001), who argue that the PhD is the award directed towards the development of professional researchers, while professional doctorates are directed towards providing research skills to career professionals.

1.2. Learning about doctoral study

These developments point to the need for a better understanding of doctoral study – what it is, how it is successfully completed and how it is best supported at the institutional and systemic levels. This understanding requires an appreciation of the importance to doctoral study of effective processes of socialisation to a research culture and its knowledge base, together with the development of a thorough grounding in the nature of disciplinary 'know-how' and of the ways in which it is acquired in its disciplinary setting. It also requires a heightened awareness of the nature of the conscious and unconscious learning processes that take place as a doctoral student is socialised into a research culture.

The focus in this book is upon the typical approaches adopted by students to doing the work of the doctorate. It examines the fundamental social phenomena of broad disciplinary settings, such as features of formal and informal communication, ways of marking out intellectual territory and of gaining acceptance, and processes for learning how to write in the expected style and to use the appropriate conventions. In this way, the examination of doctoral study presented in this book addresses many of the practical issues and concerns of supervisors and of doctoral students that have been highlighted in the relevant literature for nearly two decades. It also asserts a particular theoretical perspective that is essential to understanding the achievement represented by the attainment of a doctoral award.

2. METHODS OF STUDY

2.1. The First Stage

The primary sources informing this book are two empirical research projects supported by grants from the Australian Government, one from the Evaluations and Investigations Program, awarded in 1992, and one from the Committee for the Advancement of University Teaching, awarded in 1994. The purpose of the projects funded was to find out about the learning processes of doctoral students in particular disciplinary settings. The aim was to shed light on supervision processes, the nature of learning at the doctoral level, and the influence of the knowledge base upon what is learned. More than 120 academic staff and a corresponding number of doctoral students from three Australian universities were interviewed between 1993 and 1997. The majority of participants were interviewed more than once, in progressive stages of the study. Participants represented a wide range of disciplinary groupings and specialisms, including agriculture, microbiology, genetics, electronic engineering, psychology, education, economics, politics, accountancy, history and philosophy. The universities comprised one long-established research university, one small rural university, and one large metropolitan university typical of those that amalgamated with colleges of advanced education during the 1980s. The international system of doctoral thesis examination, together with the highly consistent findings of researchers in the United Kingdom and Europe, make these data and their interpretation applicable to the contexts of many countries.

In the first stage of the investigation, the nature of the learning experience for students during doctoral study in each of the broad disciplinary settings of the sciences, the humanities and the social sciences was examined. To that end, the students described in detail their learning experiences and the related issues, claims and concerns. In separate interviews, supervisors described in detail their values, together with their experiences of facilitating and supporting doctoral students' learning. A range of documentation produced by departments and supervisors for doctoral students was also examined.

2.2. The Second Stage

The second stage of the investigation involved a close examination of differentiation in disciplinary dialects. Twenty-six successful doctoral theses from the three partici-pating universities were analysed in terms of their writing styles, formats and con-ventions, using a conceptual framework derived from Becher (1989a) and informed by Bazerman's (1988) theory of systemic functional linguistics. More detail about the methodology, data collection and analyses is provided later in this chapter.

During the early stages of data collection, one supervisor described the notion of 'know-how' in doctoral research as 'learning the rules of the game'. The idea that doc-toral study is in a sense a game, or a meaningful social setting with rules, conventions, traditions and tangible outcomes, seemed bizarre at first. As the project developed, the notion could not be ignored because so many of the doctoral students interviewed appeared to identify very strongly with it. Looking more closely at the socialisation process that takes place during doctoral candidature, it became obvious that, while some disciplinary norms (values, traditions, conventions, symbols) are explicit ele-ments in the research environment, others are not, and so, in fact, doctoral study does resemble the combination of written and unwritten rules in any complex game.

3. KEY THEMES

The underlying argument set out here is that the nature of induction to disciplinary communities, and more broadly, the experience of doctoral study, can be better understood when related to the disciplinary culture in which it occurs. Culture is a complex phenomenon, no less so when the realm is academe and the inhabitants are intellectually sophisticated and socially diverse. The intellectual and cultural climates in which doctorates are achieved are dynamic, elaborate, ritualistic and ideologically rich. Because they are contextualised in communities, they are also normative. These are not features that are separate from or incidental to the achieve-ment and expression of knowledge outputs at the level of the doctorate. They are, therefore, elements that deserve to be explored in the intellectual and social contexts of doctoral programs.

The first theme concerns the nature of the intellectual and cultural climates for doctoral study, and what the related social settings contribute to the learning out-comes of doctoral students. Within this theme, the complexities of disciplinarity in epistemological and social terms are seminal. The second theme concerns how doctoral students might come to learn these complexities and understand their implications for knowledge production. It is, therefore, important to identify key disciplinary norms that constrain the ways in which doctoral students may go about making knowledge. It is important also to examine in some detail the social processes by which disciplinary norms are perceived and assimilated because these processes forge academic identities. The third theme concerns the role of supervision in facilitating the processes by which disciplinary norms are perceived and assimilated and, complementing supervision, the roles of others in the setting who play a part in the doctoral experience. The fourth theme concerns the mechanisms by which these norms are expressed in language and the extent to which the texts produced by doctoral students reflect the assimilation of disciplinary norms. A final theme concerns the important distinction between the PhD and other kinds of professional and coursework doctorates that have proliferated globally.

4. HOW THIS BOOK IS SET OUT

4.1. Part 1 – Knowledge-Making in Doctoral Programs

The book has three main parts. The first of these sets out the conceptual perspective and explains the theoretical framework in relation to key themes in the literature over two decades. Since the theoretical underpinnings of research training have been given little attention in the relevant literature, at least in the epistemological sense, and because higher education policy has largely ignored their importance, these underpinnings are given considerable attention. This conceptual perspective advanced, which treats doctoral study as a process of induction to a disciplinary culture or subculture (see also, Parry and Hayden, 1994, 1999; Parry, 1997), is set out later in this chapter.

Chapter 2 develops the theoretical framework underpinning the idea of doctoral study as a socialising agent to disciplinary and sub-disciplinary cultures. This chapter sets doctoral study in its proper context as a university degree that has evolved and developed over time, according to student demand, higher education imperatives and funding priorities. An account of broader international themes concerning doctoral programs and supervision is provided.

Chapter 3 identifies distinctive features of disciplinary culture and sets the context for understanding the learning processes involved at the doctoral level. It provides an explanation of the processes of induction to disciplinary cultures, and draws together related literature on learning for the purposes of informing an understanding of both conscious and unconscious learning processes.

4.2. Part 2 – Learning in Knowledge-Making Cultures

The second part of this book reports the findings of the two major research projects that inform this book. There is in this part a particular focus on the epistemologically-based nature of knowledge cultures as experienced by doctoral students and supervisors. This part of the book concludes by comparing and contrasting approaches to thesis writing in order to highlight the existence of epistemologically-distinctive ways

of reporting new knowledge that doctoral students must learn to master in the course
of candidature.

Chapter 4, which is centred on the academic enterprise as carried out in a university
department, concentrates on the practical strategies utilised by supervisors and
orchestrated by academic departments to facilitate the induction of their doctoral
students to particular sub-disciplines or fields that are cosmopolitan in nature. The
discussion draws on the experiences of students and of supervisors and highlights
explicit opportunities for learning and assimilating obvious disciplinary norms.

Chapter 5 explores the practical aspects of the social setting for doctoral study,
addressing in turn the typical work activities, patterns of formal and informal
communication among members of disciplines, modes of authorship, citation and
acknowledgement practices, and the socialising influence of key disciplinary players.
This chapter highlights key elements of the social setting for research, distinguishing
clearly between the organisational culture of the department, which is geared towards
the academic enterprise, and the broader disciplinary arena, which is an international
and outward-looking society of scholars.

Chapter 6 explores the nature of disciplinary 'know-how' and examines how
students come to develop it. Focusing on both formal and informal modes of scholarly
communication, different stages of candidature in various disciplinary settings
are discussed, and socialising influences and induction processes are identified.
A special feature of this chapter is that it distinguishes between conscious or explicit
learning opportunities, and unconscious or tacit learning opportunities, indicating the
contribution of each to the induction process.

Chapter 7 analyses epistemologically-based writing conventions found in doctoral
theses from different disciplines. It explains the values, conventions and traditions
that shape the format of the thesis, the overarching structure of argument, how
ideas are linked to form an argument, conventions for citation and critique, and the
strategies by which field-specific tacit knowledge is asserted. This chapter provides
concrete evidence, in the form of thesis texts, of disciplinary conventions and
traditions, of norms and counter norms, and of both conscious and tacit learning
processes. As with socially-based learning of any sort, the tacit dimension of learning plays a significant role and the implications of this dimension for higher learning
and the production of new knowledge are discussed.

4.3. Part 3 – Foundations and New Horizons

The third part of the book examines the epistemological foundations of doctoral study
and its implications both for policy makers and for practice. As the idea of the knowledge society takes root in the Western world, it becomes increasingly important to
ensure that doctoral study serves the needs of its various stakeholders – students,
researchers, employers, industry, commerce and the community. At the same time,
doctoral study must provide relevant and appropriate research skills and training to a
highly diverse student market. Taking into account the increasingly applied and transdisciplinary nature of the research enterprise, the changing nature of doctoral study
is examined.

Chapter 8 addresses implications for policy development. Those implications that concern institutional positioning are discussed and the issues facing organised research units administering doctoral candidature are also addressed. Implications for academic departments are addressed in their turn. There is also a discussion of the implications of the findings for doctoral students and for their supervisors. The discussion is framed in terms of socialisation and induction issues, concentrating on individual and collective responsibilities in the manufacture and reporting of knowledge in the academy. The importance of disciplinary differentiation is addressed within the current climate of increasing transdisciplinarity and applied research.

Chapter 9 reviews the range of organisational settings for doctoral candidature and the nature of the learning processes that take place in these settings. Accepting that doctoral research fits comfortably with the notion of communities of practice, the role and impact of tacit learning as a key element in learning at the doctoral level is discussed, drawing upon recent developments in the field of cognitive psychology and adult learning, and taking into account the increasingly professional and vocational nature of the research enterprise. This chapter reviews issues in the appropriateness and effectiveness of doctoral study programs and underlines the importance of a disciplinary grounding to successful knowledge production and reproduction. The chapter presents some constructive ideas about how students, supervisors, departments and universities might enhance and improve their doctoral study programs.

5. CONCEPTUAL PERSPECTIVE

The conceptual perspective used in this book draws upon an extensive literature on academic cultures and their organisation (Becher, 1981, 1987a, 1989a; Biglan, 1973a, 1973b; Clark, 1963, 1980, 1983; Crane, 1972; King and Brownell, 1966; Light, 1974; Lodahl and Gordon, 1972; and many others). It recognises that Tony Becher's articulation of academic 'tribes and territories', a landmark overview of academe first published in 1989 remains familiar today, drawing as it does on epistemology as the pillar of knowledge production. In this perspective, higher education, which typically comprises a nationally-unified system of universities of varying sizes, is viewed as embodying a diverse collection of disciplinary fields, each with its own sense of community, characteristic modes of enquiry, communication networks, bodies of scholarly traditions and sets of values, beliefs and conceptual structures. Each of these disciplinary fields has its cultural and epistemological foundations in a single disciplinary core (Becher, 1987a, 1989a; Clark, 1983), though specialisms within a parent discipline may be highly differentiated (Becher, 1990a). In most universities these fields are firmly but not restrictively bounded by academic departments, each of which contains one or more disciplinary and organisational cultures.

The setting for doctoral study may be considered from different viewpoints. In early studies, doctoral study was considered from an individual or interpersonal viewpoint (see, for example, Connell, 1985; Welsh, 1979, 1981). More recently, doctoral study is seen as vital to the formation of an academic life, so the tension between organisational and disciplinary cultures was emphasised (see, for example, Clark, 1983; Harman, 1988, Becher, Henkel and Kogan, 1994; Henkel 2000). Doctoral study may also be viewed in terms of the cultural and epistemological characteristics of the discipline (Becher, 1989a; Clark, 1980), and this is the focus of the present investigation. A department's disciplinary culture derives from membership by the staff of specialised academic communities that are national and international in their spheres of influence, through which knowledge in the discipline is perpetuated and advanced.

Another perspective evident in recent publications equates doctoral study with training, with an emphasis on the importance of the credential; indeed doctoral education and doctoral training are widely used as interchangeable terms in much of the recent literature. This perspective emphasizes the organization of research culture and subdues the epistemological core governing the rules for making and reporting new knowledge. In this vein, Lave and Wenger's (1991) account of 'communities of practice', and of learning to move from novice to expert in relation to them, provides a useful lens.

There is, however, a danger of overlooking the importance of the epistemological core, as Nowotny, Scott and Gibbons (2001:179) come close to doing. They contentiously describe the "epistemological core (as) empty – or . . . more accurately, crowded and heterogenous". Their latter description fits well with several related schools of thought (see, for example, Delanty 2001; Latour 1987) in which the nature of knowledge is being reconceptualised. It fits particularly well with the notion of highly diversified epistemic cultures described by Knorr-Cetina (1999).

5.1. Induction and Supervision

The disciplinary culture of an academic department or research centre is of special importance to the doctoral student trying to make sense of the uncertainty and diversity within the epistemological core supporting a field of knowledge. It provides an immediate research community and environment that is subject to particular disciplinary influences. Furthermore, the connection provided by the department to broader disciplinary networks is critical to the attainment of the doctorate because identification with a core discipline and with its typical activities, methods, values, traditions, conventions and social structures is essential to shaping the expression of knowledge in particular ways and to gaining acceptance by a particular knowledge community.

In the induction process in a department or research centre, the doctoral student learns how to make a contribution at an advanced intellectual and cultural level to a specific field. The process concludes when the student's thesis attains approval by experts in the field, possibly even from contrasting sub-disciplines, as representing an original or substantial contribution at doctoral level. It is through interaction with relevant scholars in specialised disciplinary fields that the doctoral student learns what constitutes an acceptable contribution within the discipline, the sub-discipline and, more specifically, within the specialism, and acquires the necessary skills and experience to gain recognition as an independent scholar. Increasingly, doctoral research is transdisciplinary, so there are tensions and competing imperatives deriving from the typical work activities, methods, values, traditions, conventions and social structures to negotiate as well as appropriate examiners to be identified. Induction in these circumstances is an even more complex achievement.

The organisational culture of a department supports the academic enterprise. It also supports the induction process by providing an administrative framework for the provision of supervision, that is, the guiding or directing of a doctoral student through the doctoral program. The organisational structure of the department also exhibits the central features of Lave and Wenger's (1991) "communities of practice" in which the novice learns to become an expert from practising members of the community.

Supervision, which is the mainstay of teaching at the doctoral level, involves the supervisor acting as a mentor, guide or adviser to an individual seeking to be recognised as an independent scholar capable of advancing knowledge in a specialised area within a discipline. Becoming a legitimate member of a community of practice can in some cases take place with only a minimal amount of supervision, but the role of the supervisor as mentor, guide or adviser is, in most instances, critical to its success.

5.2. Socialisation to Academic Cultures

The idea of socialisation to academic cultures is not new, though much recent research on the topic simply advances an idea first signalled by the anthropologist, Clifford Geertz (1983: 159-60) of the importance of socialisation to the academic professions:

My point is that "the natives' " notions about maturation (and postmaturation) in the various fields, together with the anxieties and expectations those notions induce, shape much of what any given one is like, "mentally," from inside. They give a distinctive, life-cycle, age-structure tone to it, a structure of hope, fear, desire, and disappointment that permeates the whole of it and that ought to be, as it has for Pueblo Indians and Andaman Pygmies but not for chemists or philosophers, looked into.

In all societies there are rites of passage, tests of maturity and competence, which individuals must endure in order to demonstrate acceptability to their elders. Through these rites of passage the norms and ideologies of the society are maintained and advanced. Academics, like Pueblo Indians and Andaman Pygmies, constitute tribes and inhabit territories, as Becher (1989a) has shown, but their nativeness is not naive and their rites are not wholly symbolic. In the society of universities, doctoral study represents the universally accepted rite of passage. The doctorate is the test of intellectual and cultural maturity and competence. Its achievement signifies exemplary knowledge of an academic field, arising from a sophisticated understanding of how that knowledge is shaped. In relation to the social nature of academic cultures, Thomas Kuhn (1970: 11) observed:

Men whose research is based on shared paradigms are committed to their same rules and standards for scientific practice. That commitment and the apparent consensus it produces are prerequisites for normal science, i.e., for the genesis and continuation of a particular research tradition.

An essential feature of socialisation to academic cultures and of induction to them is that academic communities and the knowledge bases they share are continually evolving: they are by no means static entities. An exploration of socialisation and induction processes in academic cultures must take account of the contradictions of fragmentation and integration that, as Clark, (1980, 1987b), and more recently, Gibbons and colleagues (1994) have observed, are central to the evolution and development of academic professions. Essential elements of these contradictions are determined by disciplinary fields and cultures, for these provide the fabric from which scholarly pursuit is cut. Knowledge of the nature of scholarly work, differing as it does across broad disciplinary groupings, is essential to an understanding of induction, because doctoral students are immersed during candidature in research cultures that are deeply rooted in disciplinary traditions that, though complex, must be mastered.

Central to the idea of induction to academic disciplines are the concepts of culture and socialisation: these are defining concepts in the development of a disciplinary identity, for disciplinary cultures are maintained and perpetuated by means of identification with disciplinary norms and ideologies (Becher, 1987a; Clark, 1980; Gerholm, 1990, Henkel, (2000).

5.3. Disciplinary Identity

Some forty five years ago Becker and Carper (1956:298) asserted that graduate students develop disciplinary identities as a result both of the structural components of graduate study programs and of the social-psychological influences of socialising with colleagues in their fields. In the intervening period, the relevant literature has addressed aspects of doctoral programs and the nature of disciplinary settings for graduate research, including their powerful cultures. The importance of socialisation and induction during doctoral study has been addressed in the related literature, although its impact on policy development has some way to go.

However, it was Clark (1983) who linked the concept of academic identity to distinct disciplines. The importance of developing a disciplinary identity during doctoral study has been identified be several researchers (Parry and Hayden, 1994; Parry, Atkinson and Delamont, 1994, Delamont, Atkinson and Parry, 1997, 2000). More recently, Henkel (2000:19) explored the concept of academic identity in some detail, distinguishing "the complexities and tensions inherent in two major sources of identity [the discipline and the institution or enterprise], one local, visible and tangible, the other cosmopolitan, largely invisible and disembedded."

This examination of doctoral study emphasises how doctoral students develop a disciplinary identity in settings of inherent cultural tension, as described by Henkel (2000). The concept of disciplinary identity used here draws upon the broader definition of academic identity that has been elegantly explained by Clark (1989), Becher (1989a) and Henkel, (2000). The examination is also grounded in long-accepted practices of appraisal at the doctoral level: the condition that doctoral theses are examined by scholars drawn from a cosmopolitan, disembedded disciplinary society, even though it is at enterprise or institutional level that doctoral study is actually carried out. Doctoral study concerns the means by which doctoral students develop exemplary knowledge of an academic field and of the symbolic parameters that shape expression of that knowledge. While Henkel

(2000:164) notes that policy change in higher education in the United Kingdom has induced productivity and performativity "shaped by norms and modes of working derived from the sciences", there can be no doubt that doctorates remain essentially value-laden and discipline-specific.

5.4. Language and Socialisation

While it is with induction that this book is primarily concerned, considerable attention is drawn to the role of language in induction and socialisation. Knowledge outputs fulfil the mission of academic cultures, and language is the vehicle through which the knowledge outputs are expressed, shaped by the predispositions arising from "different rhetorical situations, aiming at different rhetorical goals and embodying different assumptions about knowledge, nature, and communication" (Bazerman, 1987:125). We shall see by examining informal and formal modes of scholarly communication, and by analysis of doctoral thesis text structure, how induction is the means by which these predispositions are formed.

In examining at close quarters the characteristics of discipline-specific language, as exemplified in doctoral texts, the discussion is informed by empirical findings reported elsewhere (Parry, 1998) in which the epistemological features of discipline-specific dialects are examined. Various discipline-specific characteristics and features of academic dialects are examined, underlining how and why doctoral students identify with them. In doing so, we can observe fundamental aspects of disciplinary lore and culture, perpetuating themselves through the take-up of dialects by new academic recruits.

6. POLICY IMPLICATIONS

There are clear implications for supporting effectively the doctoral study experience in different disciplinary settings. The findings reported in this book should inform systemic policy considerations relating to issues such as progress and attrition rates. They should inform policy at the institutional level so that doctoral programs can be better supported in specific ways that suit the student profile and institutional context. They should also inform policy at the level of the academic department, providing guidance about the kinds of opportunities for effective socialisation to particular disciplines. In addition, the findings reported here should provide insights for supervisors as to how they might better support the doctoral research enterprises for which they take guardianship.

This book does not seek to throw light on the nature of, or issues in, doctoral examination. National systems of doctoral examination (and defence) vary considerably, even though there are efforts currently underway at the behest of the European Commission to attempt to address this inconsistency within the European Union. To date, the central issue in doctoral examination lies in the setting of standards for the award. Since, in most developed countries, universities are self-accrediting institutions, there is a tension between accepting preordained standards for the award and maintaining the freedom to determine award standards at the institutional level. The largely independent, external processes of examination provide, however, a sound framework for consistent standards in doctoral awards within particular fields of study. While there is to date no national system that requires the training of examiners in standards, some universities in Australia and the United Kingdom appear to have set training programs in place to complement the institutional instructions sent out to prospective examiners of doctoral theses. The highly dynamic and diverse nature of the PhD, as we shall see, makes it difficult for preordained standards to be articulated, other than informally, within scholarly networks of specific fields.

CHAPTER 2

DOCTORAL STUDY AND DISCIPLINARY LEARNING

1. THE DOCTORATE IN CONTEXT

Traditionally, completion of a doctorate has been synonymous with the attainment of the award of Doctor of Philosophy (the PhD, or at some universities the DPhil), granted normally on the basis of completion of a program of studies resulting in the submission and acceptance of a thesis that makes an original or substantial contribution to knowledge in a discipline. The term 'doctor' was in use from medieval times to connote distinction in fields of teaching and learning. The modern doctorate, however, in the form of the PhD, dates from the early nineteenth century in Germany (Simpson, 1983: 17), from where it spread to the United States, and then, in the early twentieth century, to Britain. The varying economic, cultural and political emphases in these systems produced distinctive forms of doctoral program. Whereas in the German system the award was tightly regulated by an examination system adopted across universities, in the American system no such regulation was encouraged, resulting in doctoral programs of varying requirements and standards (Simpson, 1983: 19–20). Another difference is that in the German and British systems there remained a high level of adherence to the singular importance of the thesis as the basis for the doctoral award, but in the American system far more emphasis came to be given to coursework in the early stages of PhD candidature (Bowen and Rudenstein, 1992). Colonialism resulted in the spread of these various forms of the PhD. Australia, for example, inherited the British model – the first PhD awarded by an Australian university was in 1948 (AVCC, 1990a: 8).

A doctoral program entails mainly that the student should complete the research required for submission of a thesis. During the past two decades, however, there have been several noteworthy departures from this traditional model. First, there has been a significant growth in the incidence of doctoral programs that do not lead to a PhD, but are instead focused upon the acquisition of advanced professional knowledge, often acquired through the completion of coursework together with a minor thesis, and invariably labelled as relating to a field of professional knowledge (for example, Doctor of Business Administration, Doctor of Education, and Doctor of Jurisprudence). These programs, though of increasing interest and importance, are not generally as prestigious, nor are they yet as ubiquitous, as the PhD. Bourner, Bowden and Laing (2001) have comprehensively discussed professional doctorates, which are not the focus of interest in this book. Second, there has been a discernible shift within PhD programs in favour of the inclusion of a coursework component, as in the American tradition of the PhD. Across Europe, for example, it is becoming increasingly common for elements of coursework to be included in

15

doctoral programs (Neave, 2003). In many cases, though, coursework inclusions are informal and supplementary in nature to the task of thesis completion. Henkel (2000: 161) notes, for example, that informal coursework is becoming more common in the humanities and social sciences, pointing to "a newly defined stage in postgraduate education for identity formation . . . [suggesting] the benefits of an earlier and clearer identification of, and requirement for, discipline-specific skills and understanding in postgraduate education."

The standard of the PhD award is maintained by an examination system that generally requires the involvement of external examiners, that is, examination by scholars from outside the university granting the doctoral award. In several countries – Norway, the United Kingdom and the Netherlands, an oral defence of the thesis is required in addition to an external examination. The examination process for the PhD is important in maintaining the standard and comparability of the award across national higher education systems. It is also important in contributing to the esteem of the award. Its esteem is enhanced by its "connotations outside academia" and by the fact that it is a licence to teach at a university (Phillips and Pugh, 1994: 18). Clark (1993) and Becher, Henkel and Kogan (1994) have noted, however, the relatively marginalised nature of graduate study compared with the investments made by universities in their undergraduate programs.

2. ONGOING ISSUES

Various longstanding issues affect doctoral education in most advanced higher education systems in the world. These were first systematically identified during the 1980s, as numbers in doctoral programs began to expand rapidly. They included issues related to student dissatisfaction with the quality of supervision, staff uncertainty about many aspects of their supervisory role and responsibilities, the lack of skill development programs in research and writing for higher degree students, the inadequate level of recognition of the importance of the research undertaken in universities by doctoral students, and generally poor retention and completion rates for doctoral programs (see, for example, Rudd, 1975, 1985; Welsh, 1981, 1982; Barrett, Magin and Smith, 1983; Ibrahim, McEwen and Pitalbo, 1980; Moses, 1984; Powles, 1984). A more sophisticated understanding of these issues emerged during the 1990s as additional insights were reported in the literature. Intensive investigations in Australia (AVCC, 1990a; Powles, 1989a, 1989b; Cullen, Pearson, Saha and Spear, 1994) brought definition and scale to the extent of the issues, and proposed a range of practical measures for addressing them. Comparable insights were reported in literature from Britain (Burgess, 1994; Becher, Henkel and Kogan, 1994), Canada (Holdaway, Deblois and Winchester, 1992; Holdaway 1994) and the United States (Bowen and Rudenstein, 1992).

In response to these issues, and in a context of increasing financial stringency across most public systems of higher education, the management of doctoral programs has generally become more tightly organised. There are more explicit rules and regulations; supervisory strategies are more explicit; and students from all disciplines are encouraged to publish during candidature. Henkel (2000: 158), reflecting on these changes in Britain, states that "education and training in all disciplines is now being shaped by the norms and modes of working derived from the natural sciences". What is not yet clear, however, is whether these developments are resulting in an improvement of the quality of doctoral candidature or an increase in retention and completion rates.

These issues, and the appropriateness of the various policy responses, cannot be considered without some reference to the role and importance of disciplinarity. There are important disciplinary influences on all of the issues identified. Effective supervision in the sciences, for example, may be entirely inappropriate in the social sciences and the humanities; and the conditions that generally affect the progress rates of doctoral students across these three broad disciplinary groupings are often quite different.

3. DISCIPLINARITY

Although there are many ways to categorise academic endeavour, a central focus in doctoral research must be upon the categorisation of subject matter in relation to disciplinary knowledge bases (King and Brownell, 1966) and the contrast between the broad epistemological properties of these knowledge bases (Biglan, 1973a; Kolb, 1981; Lodahl and Gordon, 1972). Other dimensions must also be taken into account. One concerns the dynamism and evolution of fields of knowledge (Kuhn, 1962; Toulmin, 1972). Another concerns the sociological characteristics of the communities that inhabit particular knowledge areas (Becker and Carper, 1956; Crane, 1972; Knorr-Cetina, 1989; Whitley, 1984). In terms of doctoral supervision, the interrelated nature of these dimensions is well documented (Becher, Henkel and Kogan, 1994; Parry and Hayden, 1994). In these studies, and in many others examining the nature of doctoral study or supervision, the knowledge base plays a critical part in shaping social factors that in turn facilitate the induction of individuals into academic disciplines.

Some theorists have classified disciplines according to the nature of their knowledge bases. Biglan (1973a), for example, drew upon Storer's (1966) hard-soft and pure-applied dimensions as ways of categorising knowledge, focusing on the nature of the subject matter of research, as reported by faculty members. Biglan distinguished between hard and soft knowledge in terms of the degree to which there was a shared inquiry paradigm, echoing Kuhn's distinction between paradigmatic and non-paradigmatic fields. He distinguished between pure and applied knowledge in terms of its concern for application to practical problems. He distinguished between life and non-life sciences according to the extent of orientation to living organisms. Knowledge for Biglan could be categorised as pure, in which case it is either hard (as in the natural sciences) or soft (as in the humanities and social sciences), or as applied, which could also hard (as in the science-based professions) or soft (as in the social professions). Not inconsistent with Biglan's categorisation was Kolb's (1981) categorisation of subject matter according to styles of intellectual enquiry. Kolb judged knowledge to be reflective and either abstract (as in the natural sciences) or concrete (as in the humanities and social sciences). Alternatively, it could be active and either abstract (as in the science-based professions) or concrete (as in the social professions). Yet another classification was derived from ranking subject areas according to perceived levels of paradigmatic development (Lodahl and Gordon, 1972). Becher's landmark study (1989a) synthesised the epistemological features of knowledge areas and linked these to the societies that inhabit them, vesting his argument with the elements of fragmentation and developmental change derived from Kuhn's (1962) conception of paradigms as dynamic and epistemologically driven. These categorisations have a high degree of consistency. Most recently, Braxton and Hargens (1996) have drawn together various knowledge categorisations and noted their internal consistency. They acknowledge that, whatever the label, these categorisations have "produced the lion's share of empirical work", making them widely accepted conceptual frameworks for understanding the epistemological foundations of fields of knowledge.

Biglan, Kolb and many other theorists have taken account of an idea advanced by Bourdieu (1988) that socialisation to academic life takes a good deal of time and is a major investment in the acquisition of highly specialised skills, competencies and attitudes. Bourdieu also argued that the expression of these qualities is through language, as many others, including Geertz (1973), have identified. Indeed the expression of discipline-specific language has identifiable dialects with features deeply rooted in the epistemologies of specific fields, as has been argued elsewhere by the author (Parry, 1998).

To understand how the nature of the knowledge base shapes learning at the doctoral level, it is essential to understand the core semiotic and social attributes of disciplines and sub-disciplines, as well as their evolutionary nature, their capacity for dynamism and change, and their current heterogenous and uncertain knowledge bases (Nowotny, Scott and Gibbons, 2001).

4. CORE ATTRIBUTES

There is value in categorising knowledge bases into broad disciplinary groupings, as Becher (1989a; 1990b: 334) has done. Categorisation by core attributes permits us to understand how individuals identify with specific disciplines. It also enables us to see how epistemological features play a central role in the traditions, conventions and values of the communities that inhabit specific knowledge bases. The development of a disciplinary identity (Parry, Atkinson and Delamont, 1994; Henkel, 2000) and mastery of core disciplinary attributes (Becher 1989a) and communication behaviours (Bazerman 1981) in knowledge communities are pivotal to success in making and reporting knowledge at the doctoral level.

According to Becher's (1989a) classification, pure science, as exemplified by physics, is described as hard-pure, reflecting the nature of its knowledge base as cumulative, atomistic and concerned with universals, quantities and simplification; resulting in discovery or explanation. The humanities, as exemplified by history, and the pure social sciences, as exemplified by anthropology, are described as soft-pure, being reiterative, holistic (organic, river-like), concerned with particulars, qualities and complication; resulting in understanding and/or interpretation. The technologies, as exemplified by mechanical engineering, are described as hard-applied, being purposive and pragmatic, producing know-how via hard knowledge and concerned with mastery of the physical environment; resulting in products and techniques. The applied social sciences, as exemplified by business studies or education, are described as soft-applied, being functional, utilitarian, producing know-how via soft knowledge, and concerned with the enhancement of professional practice; resulting in protocols and procedures.

Academic disciplines may accordingly be seen as represented in different forms of knowledge, or cultural capital, as Bourdieu (1977, 1990) has argued. They may also be seen as represented in an organising framework of cultural dispositions, or 'habitus' (Bourdieu, 1977), which accounts for the socially based nature of disciplinary communities. The inextricably linked nature of knowledge, its forms of development, and the cultures of the communities engaged in this endeavour, is neatly set out by Clark (1983: 34, 76; cited in Becher, 1987a), as follows:

The [academic] profession has long been a holding company of sorts, a secondary framework composed of persons who are objectively located in diverse fields, and who develop beliefs accordingly. . . . Around distinctive intellectual tasks, each discipline has a knowledge tradition – categories of thought – and related codes of conduct . . . there is in each field a way of life into which new members are gradually inducted.

Internally, however, these disciplinary groupings, like any cultural setting, are not static entities. They are in a state of continuous development, resulting in the creation of new sub-disciplines and specialisms. There are various related theories concerning the growth of new scientific knowledge structures. Pantin (1968) asserts a structure based upon whether the nature of inquiry is restricted, as in physics, or unrestricted, as in biology. Kuhn's (1962) theory of paradigmatic development asserts an evolutionary conception of the development of knowledge communities in which new paradigms develop in response to the inadequacies of a field's theoretical and methodological tools for inquiry. Toulmin (1972: 130) asserts a different impetus for disciplinary development, arguing that within a common scientific purpose, specialisms change as different concepts and theories are introduced independently, at different times and for different purposes.

All of these theories touch on a common feature of academic disciplines: the widely recognised elements of change, internal conflict and dynamism inherent in the fabric of constantly developing knowledge domains and their associated cultural settings. This is the feature that precipitates, within disciplines, sub-disciplines and, within them, specialisms. As many researchers have pointed out, these evolving specialisms are not easily separated from the communities inhabiting them.

Dynamism, fragmentation and change are easily observable across the disciplinary landscape. There are, for example, many cases in which specialisms exhibit characteristics that are uncharacteristic of the parent discipline, such as jurisprudence within law and cosmology within physics. There are several reasons why these contrasts occur. There may be overlaps of specialist interest, conflicting interpretations of the same phenomena, divisions of intellectual labour or identification with another discipline according to methodological or theoretical orientations. In short, fragmentation and change are necessary elements of disciplinary development and maintenance. Rather than being at odds, the concepts of disciplinary unity and fragmentation are linked by the common feature of dynamism. Indeed, by accepting that there are core elements in broad disciplinary groupings, we can better understand the nature of disciplinary evolution and change. It is also possible to better understand the current extent of epistemological boundary-crossing (see, for example, Weingart. 2000) in ever expanding transdisciplinary fields of knowledge, and the complexities that doctoral students have to learn during their apprenticeships.

5. NORMATIVE BEHAVIOUR

There is much empirical work in the field of social psychology to demonstrate that societies are normative, that is, people will, "if given the time and opportunity, frame their reactions and shape their behaviour in ways that are conventionally rational and normatively appropriate" (Lerner, 1978). The normative nature of scientific societies is well documented. Merton's (1957; 1973) identification of the norms of science highlights important attitudes and beliefs to which doctoral students need to be acculturated. Other contributions to the literature, emphasising knowledge as contextually bound and portrayed by the knower, offer a similar perspective (Mitroff, 1974; Storer, 1966). Mitroff's (1974), account is especially valuable because it provides evidence that the norms that are formally and explicitly encouraged do not always represent the actual range of norms that operate in any given setting. In fact, academics frequently behave in ways that are counter to the officially sanctioned rules of conduct; they conform to inexplicit, sometimes covert, counter-norms. In societies that make and reproduce knowledge, counter-norms are no less important in shaping the dispositions of individuals in the culture than are those that are more obvious. That doctoral students have to learn to conform to explicit norms and inexplicit counter-norms in making and reproducing knowledge tells us how important it is for them to be aware of both sets of norms in their research settings, and to seek them out wherever possible. It also tells us that they must have adequate access to their disciplinary communities – to have opportunities for learning from the 'habitus' of their disciplinary setting, and that this need is more pressing in fields whose epistemological bases cross disciplines and sub-disciplines.

There exists a broad literature on the normative aspects of knowledge making and knowledge reproduction in different disciplinary fields. Theorists, such as Merton (1957, 1973), whose views of knowledge development are based on epistemological and institutional boundaries, are sometimes referred to as espousing an internalist standpoint. Other researchers examine how knowledge is constructed within the different social settings of science disciplines (Amann and Knorr-Cetina, 1989; Knorr-Cetina, 1981a; Latour and Woolgar, 1979; Woolgar, 1982). In these studies, the working lives of academics are examined in terms of the underlying sociological and philosophical phenomena underpinning the construction of knowledge. In this vein, there are studies that contrast the world views of different disciplines in relation to a broad range of influences, including extrinsic social and political forces (Knorr-Cetina, 1981b; Mendelsohn, Weingart and Whitley, 1977; Lammers, 1974). In some of these studies there is a strong emphasis on internal conflict in, and on the diversity of, disciplinary cultures, echoing Kuhn's (1962) theory of the revolutionary emergence of divergent paradigms within disciplines.

Building on this literature are studies, for example, Whitley (1980), that examine in detail how attitudes and patterns of research are shaped by the intellectual tools of a particular field. While internalist theories of change seem to account for much of the differentiation among disciplines, the position is by no means resolved, as Becher, Henkel and Kogan (1994) point out. The range of norms operating in a given disciplinary setting is especially complex and does not relate to epistemology alone. Not only are political, historical and intellectual contexts relevant to the expression of disciplinary culture in different knowledge communities, but also the norms are not always clear and unambiguous. Their importance has been signalled in the literature, but the complexities of what is learned and how learning takes place have been given little empirical attention, except perhaps by Bazerman (1995).

Becher (1989a) has made a major contribution to our understanding of disciplinary norms and the processes of socialisation to them. Concerned with the way in which academic cultures are epistemologically, socially and contextually bounded, he examined a wide range of dimensions of academic life. His work addressed the traditions and taboos, intellectual territories, and internal and external boundaries of disciplines; their competitiveness and the nature of their hierarchies; their intellectual fashions, the nature of exemplars and their myths; their hidden assumptions, processes of initiation and the nature of their relationships with the laity; their patterns of communication and publication; and their conventions for etiquette in the expression of knowledge. These aspects of academic life and disciplinary culture are socially oriented and cognitively (epistemologically and contextually) bound. Since the period of doctoral candidature is the primary socialising process to academic disciplines and their specialised fields, it is during this time that disciplinary norms are learned and operationalised.

An early study by Becker and Carper (1956) shows that through interactions with peers and professors in the process of doing graduate work, the norms of particular disciplines are revealed and reinforced. There are other studies of the sociological aspects of academic disciplines and their epistemological foundations (see, for example, Fleck, 1979; Kuhn, 1962; Knorr, 1977; Latour and Woolgar, 1979; Knorr-Cetina, 1981a, 1989; Woolgar, 1982; Amann and Knorr-Cetina, 1989; Becher, Henkel and Kogan, 1994; Burgess, 1994). Many of these studies examine how the nature of the knowledge base of an academic discipline constrains the ways in which knowledge is produced and how it is then reported; in turn, knowledge production and reporting, which are both sociological and cognitive phenomena, shape the nature of the knowledge base. Some of these studies portray knowledge as socially constructed – for example, Knorr (1977) – and less dependent upon an epistemological platform.

Becher (1993: 117) examines doctoral study in three disciplines (history, physics and economics) and highlights their normative qualities in social, cognitive and semiotic terms. He notes that "physics offers considerable scope for doctoral students to become part of the disciplinary research organisation by serving as junior members of a collective program;" while, in contrast, doctoral students in history "must expect to work largely on their own, carving out highly personal research careers on the same pattern as their mentors;" and that "economics lies somewhere between the two extremes, with some degree of commonality and interdependence of research activity but little of the large scale, well organised, pattern prevalent in some of the key areas of physics."

Three other studies have focused upon the normative qualities of academic disciplines and their influence upon what is learned during doctoral study. The first, a development from Becher (1993), is by Becher, Henkel and Kogan (1994), in which graduate education in three disciplines – economics, exemplifying the social sciences, history, exemplifying the humanities and physics, exemplifying the natural sciences – is examined. The focus here was on how research students are socialised by immersion in these specialised fields. The second study is by Parry and Hayden (1994), in which opportunities for socialisation to disciplinary conventions, traditions and values were explored in specialisms representing the broad disciplinary groupings of the sciences, the social sciences, the humanities and applied professional fields. The third, by Delamont, Atkinson and Parry (2000), was based on two major studies of the academic socialisation of doctoral students in social science and natural science disciplines. These three studies share two important elements. First, all three emphasise socialisation to specific disciplines. Second, each is empirical, utilising large databases involving between a hundred and three hundred higher degree supervisors and students. All three studies reported findings that were highly consistent, though two different national systems of higher education, the British and the Australian, were involved. Taken together, they constitute a sound basis for making generalisations about how doctoral students are socialised into disciplines, and the contribution of sociological, epistemological and contextual influences on learning processes.

6. THE SETTINGS FOR DISCIPLINARY CULTURES

A proper discussion of disciplinarity and socialisation to academic disciplines calls for some consideration of where the socialisation takes place. Disciplines and specialisms are abstract phenomena; they are not physical, concrete entities. Their landscape is international, and their populations are heterogenous. Though individual scholars are located within university departments or research centres, their values and traditions belong to the international disciplinary arena.

Delamont, Atkinson and Parry (2000: 3) argue that "a separation between academic subculture and organizational context is hard to sustain when subjected to close scrutiny". They also recognise, however, the importance of doctoral study as the primary opportunity for the development of a disciplinary, that is, an internationally-oriented, identity. Considering that the doctoral thesis is the single indicator of disciplinary competence, and that it is examined, except in rare cases, by scholars drawn from the international disciplinary arena, it is the discipline and its norms that are of most importance and value to doctoral students. For this reason the discipline is identified as being the intellectual society to which the doctoral student aspires, while the department, research centre or community of practice (Wenger, 1998) is the organisational unit that provides opportunities for socialisation to a specific discipline or field.

CHAPTER 3

SOCIALISATION

1. FORMING SOCIAL IDENTITIES

Delamont, Atkinson and Parry (2000: 4) write that "there is no doubt that doctoral research produces and reproduces not only knowledge, but social identities as well", and that "evidence from the international research [shows] that identities are discipline-specific". This view is shared by a number of other researchers concerned with socialisation to academic disciplines, including Becher, (1989b); Kogan (1988) in history; Becher (1990a); Youll (1988) in physics; and Brennan and Henkel (1988) in economics. A feature of much of the related research is that it recognises the socialising power of disciplinary communities as a means of perpetuating and advancing disciplinary cultures. In a book edited by Clark (1993), a cross-national perspective on disciplinary identities is provided in which physics represents the sciences, economics represents the social sciences and history represents the humanities. Though the disciplinary identities in these fields are marked, highly differentiated systemic and organisational conditions affecting graduate research are also reported – from Germany (Gellert, 1993a, 1993b), the United States (Gumport, 1993a, 1993b), the United Kingdom (Becher, 1993; Henkel and Kogan, 1994), France (Neave, 1993; Neave and Edelstein, 1993), and Japan (Ushiogi, 1993; Kawashima and Maruyama, 1993).

The development of distinct social identities in academe is a complex phenomenon. As Henkel (2000) argues, there are many aspects of the socialising process that need to be considered. In the related literature, attention is given to a range of factors thought to influence the nature and quality of the doctoral study experience. In some studies, the role of the department and the supervisor are emphasised (see, for example, Blackburn, Chapman and Cameron, 1981; Elton and Pope, 1989; Hartnett and Katz, 1977; Moses, 1984, 1992; Phillips and Pugh, 1987; Rudd, 1985; Welsh, 1979); notions of collegiality and mentorship are identified as contributing to successful socialisation. The problem of student isolation, whether intellectual, geographic or emotional, during doctoral candidature is also widely reported (see, for example, Moses, 1984, 1992; Powles, 1988a, 1989a; Rudd, 1985; Wason, 1974; Welsh, 1979). In several studies isolation is held to account for poor completion rates and times. This problem is more likely to occur in social science fields (Hockey, 1991), where the typical student profile comprises a relatively high proportion of students who are mature age, part-time, female, faced with competing work and family responsibilities, or a combination of these (Powles, 1989b). Related to isolation, different scenarios of 'connectedness' have been described, with Delamont and Eggleston (1983) pointing out the tension between originality and intellectual loneliness during candidature. All of these studies suggest that there is a fine balanced between the emotional, intellectual and epistemological aspects of socialisation to a disciplinary culture during doctoral candidature.

In a study of doctoral research in the social sciences, Burgess, Pole and Hockey (1994) examine processes of formal and informal socialisation involving students and their supervisors in the first year of candidature. They note particularly that an effective introduction to a research culture, and to a culture of support and completion, are critical to success. In a complementary study, Parry, Atkinson and Delamont (1994) find that disciplinary identity transcends the organisational culture of academic departments. There is a strong social orientation in the learning process that relates to the disciplinary 'core' that the student adopts as their research focus rather than to the social setting of the department.

2. CHANGING ORGANISATIONAL SETTINGS FOR DOCTORAL RESEARCH

Doctoral research is largely carried out in or in academic departments, though recent developments in higher education have meant that research centres dislocated from academic departments and sometimes universities are increasingly common, especially in science. In turn, the nature of knowledge production is also changing. Whereas knowledge was once largely embodied within the walls of universities, we now freely use terms such as "the knowledge society" to convey the notion that knowledge is community based, freed from the political and clerical confines of the past. Doctoral study has not been immune to these changes.

In a comparative study of doctoral programs in Australia, the United Kingdom, Canada and the United States, Noble (1994) noted the lack of homogeneity and equivalence across national systems – an issue later picked up for continental Europe in the Bologna Process (Burgen, 2003). However, not until quite recently (see, for example, Neumann, 2002) have researchers turned their attention to the lack of homogeneity *within* national systems.

For the most part, the lack of homogeneity is largely the product of the comparatively different histories of higher education sectors, but it can also to some extent be ascribed to the increasingly applied and transdisciplinary nature of the research enterprise (Brennan and Shah, 2000; Symes, 1999) in a climate where many national systems are spending relatively less on higher education, thus passing the costs on to students (see, for example, Long and Hayden, 2001). Under these conditions, students are more vocationally oriented and (see, for example, Bourner, Bowden and Laing (2001), so that doctoral students are now more likely than ever before to pursue careers outside academe (see also Turner, 2000).

Not surprisingly, coursework and professional doctorates have proliferated in some countries, notably the United Kingdom and Australia. The inclusion of coursework within research doctorates along the lines of the US system, is also increasingly evident in Europe, the United Kingdom and Australia (Bourner and Laing, 2000). Furthermore, the research enterprise is increasingly supported by links with industry, commerce and the community. Doctoral research may be carried out in independent research centres, and it is not uncommon for students to have at least one supervisor from outside the university – for example, from industry. All these fairly recent developments in the settings for doctoral research have implications for learning at the doctoral level, and for the student's proximity and access to disciplinary cultures and appropriate learning opportunities.

They have also had an impact on the organisational settings for doctoral study and the socially-oriented nature of the learning that takes place. Herd, McWilliam and James (2000) see in these developments that universities are being affected by industry and government relationships, and they argue that the notion of doctoral education is a contested term. In this vein Delanty (2001: 152) argues that "universities are losing their role as the sole site of knowledge production, for knowledge is now being produced, or at least shaped, by many other social actors". Traditional disciplinary-based departments are being replaced by more contemporary enclaves such as cultural studies, ecotourism or informatics, which are closely linked to practice and the world of work. However, commenting on the tension between established notions of disciplinarity and newer conceptions of knowledge production, such as those of Gibbons and colleagues (1994), Weingart (2000: 38) cautions that:

... the social scientist observing these processes must be warned not to take the rhetoric of competing exaggerations at face value ... Structures of knowledge production like any others reflect the fundamental distinctions, ordering categories, and social representations that are necessary to maintain the activity, to give it direction for the future by providing the memory of past achievements.

2.1. Knowledge Production in Communities of Practice

Lave and Wenger's (1991) theory of situated learning provides a useful perspective for addressing Weingart's concerns. Such a perspective also accommodates newer conceptions of knowledge production embodied by Mode 2 knowledge societies as described by Gibbons et al., (1994). Situated learning is the process by which individuals learn from relevant social actors, form identities and develop community values. Advancing this idea, Wenger (1998) describes the communities of practice in which the social learning takes place, likening them to professions and highlighting community values, identity formation, experiential learning and practice as core elements. Wenger's perspective is well suited to professional practice and to the increasingly applied nature of doctoral research. It also fits neatly with Latour's (1999) conception of doctoral education as a learning experience in which networks of academics, administrators, students, politicians, employers, professional associations and others all play a significant part. The related notions of situated learning and communities of practice are silent, though, on the nature of the knowledge base that informs professional communities of practice.

Noting the different contexts that need to be accounted for in doctoral study programmes, it is important to acknowledge that doctoral study also has an organisational context – *which in itself may be considered an administrative or organisational community of practice*. However, it is most important to use the lens of the nowledge base as informing the community of practice when it comes to knowledge production at the level of the doctorate because candidates are principally concerned with coming to understand how to contribute legitimately to a particular knowledge base and associated community of practitioners. While the organisational settings for doctoral research may be changing, with many institutions now supporting graduate schools and colleges in an effort to reduce isolation and improve completion rates and times, the nature of knowledge communities remains normative, secular and epistemologically-bound. We now turn to the explicit and the implicit nature of norms and conventions, and the complexities of learning them.

3. EXPLICIT NORMS, CONSCIOUS COGNITION

Studies building on the importance of disciplinary differentiation point to there being explicit norms, with associated explicit means of learning them, as well as norms that are inexplicit, less obvious or perhaps learned at the unconscious level. It is the explicit norms and how they are learned that is the focus of most of the research on doctoral study. Becher (1993) examined explicit disciplinary norms and explicit processes for learning them at each stage of candidature, from processes of research supervision and the nature of doctoral programs to issues such as topic choice, the role of the supervisor, formal and informal socialising opportunities, and monitoring and assessment.

Becher, Henkel and Kogan (1994) went further by providing a synoptic model of graduate education in the United Kingdom, in which the prevailing explicit norms and conditions for graduate research are examined, focusing especially on doctoral study. Within the sciences, physics and biochemistry were contrasted, as were economics and sociology within the social sciences, and history and modern languages within the humanities. Becher, Henkel and Kogan (1994) showed that there is substantial differentiation within broad disciplinary groupings as well as across them. They argued that doctoral research in science is enmeshed in the organisation of research, while, in the social sciences and humanities, the research enterprise is individualistic, so that doctoral research efforts are distinct from the work of established scholars. In these specific settings, explicit epistemological features of disciplines can be seen to shape the nature of the associated communities; through immersion in their research communities, students learn explicit norms and how to manipulate them.

Parry and Hayden (1994) identified a range of explicit processes typically put in place by academic departments to provide doctoral students with opportunities to learn specific disciplinary values, conventions, and traditions. The processes are described as discipline-specific strategies for managing each of the stages of candidature, and they include recruiting and selecting students, allocating supervisors, providing guidelines about departmental and disciplinary expectations, selecting the topic, giving advice about how to do research, meeting with students, helping students to write, maintaining a working relationship, checking progress, introducing students to scholarly networks, ensuring acceptability of the thesis, selecting examiners and providing career support. In each disciplinary grouping, these strategies are shown to a distinct consequence of their various epistemological underpinnings.

Socialisation processes in natural science and social science disciplines have been explored more recently (Delamont, Atkinson and Parry, 2000: 3–4), ranging from laboratory research in the natural sciences, through computer-based modelling, to field research in social sciences, taking into account disciplines as varied as biochemistry, physical geography, artificial intelligence, town planning, human geography and social anthropology. They described the features of 'enculturation' in these fields of knowledge, focusing on explicit disciplinary norms and conscious opportunities for learning them.

The importance of these and similar studies has been their capacity to inform policy and the strategic development of doctoral study programs across many national systems. It is now more widely recognised that opportunities for being inducted into specific sub-disciplinary fields are a necessary component of doctoral programs. It is also becoming accepted that socialising opportunities must be very specific: they must allow the doctoral student to learn the social as well as the semiotic values, traditions and conventions for making and reporting knowledge in a particular field. Against this background, Henkel's (2000) large-scale study of the development of academic identities raises a significant issue:

The doctorate had finally become institutionalised across the spectrum of disciplines as the foundation of an academic career . . . research or the production of knowledge remained central in the process of academic identity formation. It was now a more managed form of production, in more elaborately structured institutions, larger units. There were clearer and more limited time frames within which more tangible evidence of progress would be required. There were more rules and procedures. These developments were broadly in tune with developments in the organisation of the natural sciences, where unit size, collective forms of work, pooled resources and various forms of collaboration were becoming increasingly important. The time frames were also more congruent with the norms of scientific publication . . . They reflected instrumental rationality rather than the rationality of communicative action (Habermas 1984).

It can be argued further that the norms of the natural sciences are not necessarily adaptable to the epistemologies of specialised fields in other disciplines. Henkel's argument highlights two key issues. The first is that there is a widely held assumption that the norms for collective areas of research endeavour can be applied successfully across all disciplines, including those that are highly individualistic, such as in the humanities and social sciences. The second is that the breadth and impact of disciplinary norms is not well enough understood. The latter issue is addressed in semiotic terms in an analysis of thesis writing (Parry 1998), which shows that not all disciplinary conventions, traditions and values are explicit. But tacit norms are not limited to language. The inexplicit nature of many disciplinary norms and the tacit means by which they are learned can be educed in many of the empirical studies of doctoral programs and supervision already referred to in this chapter. But it is necessary also to consider discipline-specific discourse because knowledge *reproduction* is an accessible vehicle for the examination of both explicit and inexplicit conventions and values in knowledge *production*.

4. INEXPLICIT NORMS, UNCONSCIOUS COGNITION

Because academic cultures are epistemologically conditioned, language is an essential means for the expression of those cultures. It is, therefore, surprising that there are few empirical studies of writing at the doctoral level, especially since there is a broad literature on the characteristics of academic writing in other scholarly texts. Useful findings about how academic writing meets the expectations of the academic audience (Swales, 1983) have added to our understanding of explicit steps in constructing scientific journal article introductions. Swales argues (1983:196) that this formalised strategy is actually a plea for acceptance, made in a way that is acceptable in the discipline, and he uses the jigsaw metaphor to illustrate the necessity of including all steps in the strategy. He claims that the strategy is especially relevant to novice writers such as graduate students and students from non-English speaking backgrounds. However, Swales does not extend his theory to take account of disciplinary differentiation.

The idea of writers having to meet the expectations of highly differentiated audiences is advanced by Bazerman (1981), who provides striking contrasts between three exemplary journal articles and thereby highlights the different purposes of writing in the science, social science and humanities disciplines. Bazerman crystallises differences in writing conventions as attributable to the nature of knowledge and its boundaries; the traditions for relating new knowledge to existing literature; the extent to which language is penetrable to the outsider; the nature of the technical terminology; the need for methodological and theoretical justification; and conventions for the tone and style of the writing in relation to how new knowledge claims are made. Bazerman's (1981) study is especially important because it provides empirical evidence of the key epistemological and cultural conventions used to express disciplinary knowledge, that is, knowledge of a kind essential to aspiring doctoral students.

Becher (1987b) describes in practical terms the ways in which steps are taken in constructing an argument that meets disciplinary expectations. His argument is that, in science, the writer knows the rules and that there will be an accumulation of components to the argument, which is built from 'the bottom up'. Writing in the humanities is more interpretive and explanations have to be constructed in relation to a vision of the whole argument, which can then be argued from 'the top down'. Further evidence of explicit differentiation in the conventional forms of communicating knowledge, both spoken and written, is amply provided in literature about the cultural influences on language conventions, such as characteristic patterns and forms of publication.

There is evidence that there are formal and informal differences in communication amongst scholars and in the conventions for citation (Crane, 1972), that there is differentiation across disciplines in the kinds of reputational and reward incentives for citation in scholarly texts (Cronin, 1984), and that citation differences give expression to a range of discipline-specific conventions for expressing epistemological values (Becher, 1987b). Becher draws particular attention to disciplinary differentiation in the norms for making references to existing research, in acknowledgment conventions and in the conventions of politeness in referring to other authors. The importance of these findings convey the wealth of tacit knowledge contributing to disciplinary culture, whose conventions are essential elements of cultural capital for successful doctoral students.

Codification in the dialects of disciplines and their specialisms are also noted by Zuckerman and Merton (1971; 1973; 1986), who examined journal rejection rates across disciplinary settings. They argue that the substantial variation across these settings seems to arise from differences in agreement about standards of scholarship in particular fields (Zukerman and Merton, 1971: 77). These researchers (1972) suggested that learning disciplinary know-how is easier in disciplines where there is a high level of agreement about scholarly standards, a view that is consistent with the categorisation of disciplines by both Biglan (1973a; 1973b) and Kolb (1981).

5. TACIT COGNITION AND LEARNING

The discipline-specific conventions and values inherent in reporting scientific knowledge as outlined above are, for the most part, inexplicit. As Parry (1998) indicates and as others (Bazerman, 1981, 1995; Knorr-Cetina, 1981b, 1999) imply, they are also learned by tacit means. But inexplicit conventions and values and tacit means of learning them are not confined simply to the reporting of knowledge: they apply equally to its manufacture.

Further scope for consideration of inexplicit norms and of unconscious means of learning them derives from the field of cognitive psychology, where tacit cognition has long been the subject of controlled experiments (see Gilbert, 1977; Wegener and Petty, 1997; Levy, Plaks and Dwek, 1999), and is becoming better understood. There is a widely held distinction between conscious, rational, analytic, intentional learning and cognition, on the one hand, and implicit, tacit and unconscious learning and cognition, on the other. It is thought that the two systems of cognition are productively interactive in determining learning outcomes. Reber (1997) explains:

No doubt much of what we do takes place in the full light of a modulating, controlling consciousness that, as Baars (1988) has argued, displays a rather rich array of functions. On the other hand, acquiring one's natural language, becoming socialized, being able to discern that a new piece of music is indeed Mozart, vaguely sensing in the middle of a chess game that there may be something to be gained from a novel line of attack, having an intuitive sense about a productive line of research, coming to be able to anticipate the appropriate location of a stimulus in a sequence learning experiment, learning to distinguish acceptable sequences of letters from those containing violations all take place largely in the absence of awareness of the epistemic contents of mind that accompany them.

In addition to there being two modes of cognition, one conscious and the other tacit, there is empirical evidence that the tacit learning mode is socially and experientially based: "[people will] if given the time and opportunity, frame their reactions and shape their behaviour in ways that are conventionally rational and normatively appropriate (Lerner, 1987)" (cited in Hamilton, Sherman and Maddox, 1999). This mode of cognition is also bound up with the development of socially-based attitudes and values: "Our studies . . . found that when situational information was made perceptually salient, accessible and applicable, it strongly influenced dispositional attributions . . . " (Trope and Gaunt, 1999: 177).

Empirical evidence of tacit cognition raises questions that are relevant to our understanding of learning at the doctoral level, and there is still much to discover. For the field of cognitive psychology, Hamilton, Sherman and Maddox (1999: 624) note:

Our analysis leads us to the conclusion that questions about dual-processing mechanisms in social psychology are complex, and that some of the most critical questions. . . . have seldom been seriously addressed. Among these questions are what the criteria for a true dual-process model are, whether the two proposed processes operate in parallel or are sequential, whether the processes fall along a single continuum or represent different mechanisms and structures, and under what conditions one or the other process is more likely to predominate.

The tacit dimension of learning does not lend itself easily to the rules of science – of objectivity and reproducibility. Yet it is clear from existing examinations of academic writing that the complex and sophisticated level of learning that doctoral students acquire during candidature involves cognition on both conscious and tacit levels. Reber's (1997) account, and those from proponents of dual theories of cognitive processing (see Chaiken and Trope, 1999, for a recent overview) provide an empirical base informing our understanding of the social aspects of learning during doctoral study. This empirical base is important because it is highly consistent with both disciplinary differentiation and with newer organisational models of social learning, as described by Lave and Wenger (1991) and embodied in the term, "communities of practice".

5.1. A Theory of Learning

A theoretical framework that incorporates a conception of knowledge production with a conception of the learning process and of the learning setting is essential to understanding the processes of induction of doctoral students to specialised disciplinary cultures. In particular, Reber's account fits well with a socio-semiotic view of academic cultures because he distinguishes between unconscious cognition and verbalisability as a means of measuring it. Verbalisability is not the only measure of unconscious or tacit cognition, but mastery of the discourse of a field is certainly the accepted measure of disciplinary know-how at the doctoral level.

5.2. Learning Counter Norms

Knowledge cultures to which doctoral students aspire are normative; but they also exhibit counter norms. As noted earlier, Merton (1957, 1973), Storer (1966), and Mitroff (1974) identified certain attitudes and beliefs that shape behaviour in scientific circles, long known as the 'norms of science'. Other authors, most notably Becher (1989a), have identified comparable norms in other disciplines. But normative behaviour is complex and influenced by subliminal characteristics: both the set of counter norms which shape decisions about communication behaviours laden with social consequences, and the range of inexplicit, or un-conscious norms, are learned by tacit means through socialising in the research environment.

Immersed in the social setting of research, and located in the broader disciplinary arena, the doctoral student finds that norms which are formally and explicitly encouraged do not always represent the entire range of norms that operate in any given setting. It was remarked in Chapter 2 that academics frequently behave in ways that are counter to the officially sanctioned rules of conduct, which may be referred to as counter-norms. While they may seem to run counter to the norms of a discipline, they are no less important in shaping the dispositions of individuals in the culture than are those that are more obvious:

... contrary to some arguments ... in the writings of the sociologists of science ... the current norms are not the only possible norms of science. They are neither unique nor sufficient for the attainment of scientific rationality. (Mitroff, 1974: 14).

The pivotal role of social interaction with peers, colleagues, supervisors and other students during doctoral study is to provide opportunities for important counter norms to be gleaned. Gerholm (1990), for example, describes in detail the forms of tacit knowledge that a research student must acquire in order to be successful, claiming that failure to do so "is often taken as a sign of failure to have acquired the explicit knowledge itself".

To see how complex and pervasive tacit knowledge and tacit learning are in the context of specific disciplinary settings, it is useful to consider the seven interactively overlapping forms of tacit knowledge identified by Gerholm (1990). He asserts that research students must develop the ability to differentiate appropriate behaviours for certain circumstances; adopt an appropriate attitude or *savior faire* when choosing among conflicting norms, know the folklore of the field, understand the locale of the field within the broader discipline; know how an individual research project relates to others in the field; and, importantly, develop a knowledge of scientific discourses, for this constitutes the expression of knowledge. Another category of tacit learning, argues Gerholm, is the knowledge a student acquires from the research environment and from socialising with colleagues during candidature.

Putting together Gerholm's different kinds of tacit learning with what we know about disciplinary cultures, four distinct categories of tacit knowing, or savvy, emerge. These are not amenable to sharp distinctions from one another because they are interactive, overlapping and continuously evolving with the development of the disciplinary knowledge base; they are, however, immediately recognisable in the ways in which doctoral students convey mastery of the rules governing the making and reporting of knowledge.

6. DISCIPLINARY SAVVY

The first, *grounded savvy*, grounds the novice substantively in the field, but also grounds the field in relation to the broader specialism. It involves an understanding of the nature of the phenomena being researched and its place in its broader phenomenological setting. Developing grounded savvy means becoming intellectually grounded in the theoretical traditions, methods and conceptual developments of the field. It includes a clear understanding of the methodological underpinnings of phenomena, the typical ways of making knowledge in the specialism and what and how to add to its knowledge base. Grounded savvy concerns the development of substantive knowledge together with that which Gerholm (1990) has described as understanding the locale of the field within the broader discipline and knowing how an individual research project relates to others in the field. In specialisms that are essentially interpretive or paradigmatic, the development of grounded savvy may be a tortuous process.

The second is *cultural savvy*. Echoing Bourdieu's concept of cultural capital, it concerns the development of a reservoir of knowledge of the values and disciplinary aspirations of scholars in the field. It includes understanding the folklore of the field and the influence and role of exemplars, as well as an understanding of the locale of the prevalent values and aspirations of the field within the broader discipline. Cultural savvy is drawn upon constantly in engaging in scholarly activity, for it includes an understanding of the tacit knowledge of the field – knowledge that is accepted as embedded in the knowledge base, and which does not have to be justified or acknowledged because its influence is so fundamental and influential. Cultural savvy therefore embodies what Gerholm (1991) refers to as the ability to adopt an appropriate attitude when choosing among conflicting norms.

The third, *social savvy*, concerns the ability to communicate effectively with other scholars in the field, whether in formal settings or in the everyday social life of the discipline and the department. This ability involves an acute understanding of one's place in the hierarchy of the department, and of the broader specialism, and it is manifested in appropriate disciplinary behaviour, confidence and attitude. Social savvy is essential to confidently knowing how to comport oneself among other scholars in the disciplinary arena. Its development is crucial to being able to contribute to and learn from the scholarly networks, since it concerns the form of tacit learning identified by Gerholm (1991) as being the knowledge that the student acquires from socialising in the research environment.

The fourth, *discourse savvy*, concerns the expression of other forms of disciplinary savvy and it denotes a sophisticated knowledge of the conventions, traditions, politics and techniques for reporting new knowledge. Discourse savvy is the expression of knowledge in the field using a particular dialect that is shared by like-minded scholars. It is more than a mere dialect, however, because it is value-laden. It involves taking into account the nature and structure of argument, the conventions for citation, acknowledgement, praising and blaming, the technical terminology and strategic use of metaphorical language, and an appreciation of what constitutes the already accepted tacit knowledge.

6.1. Learning to be Savvy

An interdependent mix of these four kinds of disciplinary savvy needs to be achieved by the doctoral student in order to successfully achieve their award. Yet the learning involved is largely tacit, though it makes sense that, as several cognitive theorists argue, the system of unconscious or experiential learning interacts constantly with the analytical, reflective system. Consistent with Epstein and Pacini's (1999) findings, the four categories of disciplinary savvy can be seen to interact in productive ways with the analytical, conscious mode of cognition.

The inability of supervisors to articulate explicitly subtle features of disciplinarity to their students, such as the precise language features of a field, has long seemed curious. But it can be explained by Reber's (1993:64) empirical work: " . . . knowledge acquired from implicit learning procedures is knowledge that in some "raw" fashion, is always ahead of the capability of its possessor to explicate it".

Why is there so little cohesive research into tacit learning given that it can inform our understanding of learning at the doctoral level? Reber (1993: 4) explains that experimental work in the cognitive sciences and on conditioning have had very little overlap, so that learning and cognition, which are clearly richly intertwined, unfortunately tend to be reported as two distinct fields. In recent years, researchers have provided ample evidence of 'dual processing' in learning (for example, Reber, 1993, 1997; Cohen and Schooler, 1997; Chaiken and Trope, 1999), and of the interaction between the two learning processes: "Under nonexceptional circumstances, complex human behaviour is a delicate blend of the implicit and the explicit, the conscious and the unconscious." (Reber, 1993: 25).

6.2. Tacit Learning

There is considerable evidence in the related literature for the idea of tacit norms and tacit learning, a notion that Polanyi (1966:4) expressed – "We know more than we can tell" – decades ago. This view is underscored by the influential work of Mitroff (1974) and Gerholm (1990), who identify tacit norms and describe unconscious processes for learning them. Parry, Atkinson and Delamont (1994: 46) suggest how this may work in practice when they assert that, at least in anthropology, coursework learning is far less relevant to fieldwork than 'learning by doing', because much of what is learned in fieldwork is tacit and occurs at the subconscious level. Delamont and Eggleston (1983), examining intellectual isolation during doctoral study, apply Miller and Parlett's (1976: 144) description of how undergraduates play the examination system to the approaches of research students in identifying criteria for success. They recognise that individuals vary in the extent to which they identify tacit knowledge in their environments, and they identify three categories of learners among postgraduate students. There are some postgraduate students who are alert to cues to tacit knowledge in the social setting for research, and actively seek them out ('cue seekers'); there are some are aware of cues when they are presented to them ('cue-conscious'), and those who are not aware of cues to tacit knowledge at all ('cue-deaf').

Taken together, the tacit knowledge represented by the four different kinds of disciplinary savvy – grounded savvy, cultural savvy, social savvy and discourse savvy – can account for the learning achievement that is represented by the achievement of a research doctorate. They echo the notions of 'cultural capital', to which Bourdieu refers. Additionally, they account for the dual imperatives of coming to understand and manipulate counter norms as well as inexplicit ones, as Gerholm (1990: 256) described: "A career-minded researcher had better familiarise himself with the more Machiavellian rules of conduct that both Mitroff (1974) and Pierre Bourdieu . . . have found to obtain de facto among scientists"

As the doctoral student becomes aware of, and seeks opportunities for acquiring experiential, tacit learning, he or she comes to learn what Mitroff (1974) and Gerholm (1990) have termed the implicit 'rules of the game' of knowledge production in particular disciplines and subdisciplines. What is more remarkable is that the rules of the game are so dynamic against a background of the groundswell of the knowledge society. Can the task ever have been more daunting?

6.3. Learning is Multidimensional

The argument that has been advanced in this chapter shows that effective socialisation to specific disciplines can only occur when there are ample opportunities in the learning environment for learning the conventions, traditions and values of a specific knowledge field. The chapters that follow provide ample evidence of the kinds of opportunities that are appropriate to particular disciplinary settings. They also convey the nature of both conscious, analytical modes of learning and tacit, experiential modes of learning, and the ways in which these interact to produce learning outcomes. Most importantly, they describe these aspects of learning at the doctoral level in the various stages of candidature and through the formal and informal modes of communication so vital to learning in the disciplinary setting.

PART 2

LEARNING IN KNOWLEDGE-MAKING CULTURES

CHAPTER 4

INDUCTION PROCESSES

1. IMMERSION AND INDUCTION

The entry to a new intellectual culture is commonly accompanied by the need to understand and become immersed in its formal procedures as well as its informal customs. It is therefore unsurprising to find that, while many of the procedures and customs related to postgraduate work are familiar enough to be taken for granted by supervisors, for doctoral students the process of becoming a legitimate contributor to a knowledge community is a gradual one. For them, it is necessary to understand the rules by which knowledge production in a specific field is undertaken, for it is by these rules that progress is made during doctoral study, and by which the doctoral thesis is judged. The process of writing the PhD is conditioned by a set of procedures that enable the student to develop a necessary range of appropriate skills and capacities for demonstrating and adding to the knowledge of the field. Within doctoral programs there are procedures explicitly designed to induct the student into the relevant society of researchers. This chapter reviews formal procedures and practices that help the doctoral student to develop adequate skills and capacities and to learn explicit disciplinary traditions and conventions that are linked to the nature of knowledge and its production.

1.1. Starting Candidature

The ways in which knowledge production takes place in a particular discipline are influenced by the procedures that supervisors carry out in order to meet the expectations and conventions of their fields. The effects of different forms of knowledge production on the nature of doctoral programs are first encountered at the point at which a student seeks admission to candidature.

In the sciences, the frontiers of knowledge are well known to scholars and the pace of research is fast and highly competitive. The rapid pace of research means that there is an emphasis upon the fast production and reporting of results, and upon deadlines. Becher (1987a: 282) explains the dominant values:

aspiring . . . students in hard-pure subjects may choose the department in which they wish to work; they may also be able to choose their research supervisor. Beyond this, they become, in a way, employees in an enterprise that requires their collaboration and their conformity . . . their research theme is defined for them by their research supervisor, who is likely to locate it within an area in which the supervisor and other researchers are engaged.

A postgraduate co-ordinator from biochemistry illustrated the nature of the scientific imperative when he reported: " . . . we go through troughs where things go wrong and things go badly and the research gets out of date. I don't know, that's the

only pressure I feel as a supervisor . . . that the project is progressing sufficiently."
Science students become part of a collaborative effort in which the norm is to work
towards the first priority of the research team: knowledge outputs. An important outcome of this emphasis is that in many science fields, part-time candidature cannot
easily be accommodated, and students without scholarships are advised that there
will not be time for them to earn funds outside their doctoral programs.

Those seeking admission to candidature experience rather different values in
fields where knowledge is individualistic and uncertain, such as in the humanities.
There is relatively little pressure to produce results at a rapid pace and little competition for intellectual territory. Becher (1987a: 282) crystallises the values here, too:

In soft-pure subjects . . . students are able to negotiate not only their departments and supervisors but also
the themes of their research. Far from being regarded as employees, they are treated like self-employed
persons or individuals of independent means . . . .." Part-time candidature is relatively more common as a
result of the greater autonomy of students, and completion times tend to be longer.

1.2. Topic Selection

During the introductory stage of PhD candidature, there are different procedures in
place that serve to immerse the student in the prevailing research culture and to give him
or her first-hand experience of the rules of knowledge production. These rules are governed by norms that the students soon learn. The student is expected to learn by doing,
and the rules of the research societies are themselves the rules for the doctoral student.
The procedure for selecting the PhD topic exemplifies the way in which students work
towards an understanding of the disciplinary parameters of knowledge production.

In the social sciences, topic selection is very much open to negotiation between the
student and supervisor, reflecting both the likelihood of highly individualised research
and the platform of common technical and theoretical tools evident in particular fields.
While the procedure is designed to ensure that the topic is appropriate to doctoral
study, there are certain norms of knowledge production with which students have to
come to terms. In economics, students need to learn field-specific skills that are
versatile and enduring. One supervisor explained: "We tend to encourage PhD topics
that can be tailored to what's required for three years' work. Usually the student wants
to take on more, but we know that they'll never work in the same field after they finish
. . . it's more important to have skills that they can apply afterwards to other situations."
Tailoring the topic is also important because students need to maintain a specific focus,
especially when they are working from large databases, which is typical in the social
sciences. Another economics supervisor explained: " . . . sometimes [students] have
trouble. They know vaguely the topic, but it often needs to be pruned – narrowed down
or broadened out, depending upon the scope."

In science, the production-line ethos requires that doctoral students can only
embark on those research topics for which the necessary resources can be found. The
boundaries of knowledge are well-known, so new research questions are constrained
by established research directions. The usual procedure is to give students the clear
expectation that topic selection will be in line with the activities and facilities and
existing research undertakings in the department, even though some areas may offer a

little more choice than others. A chemist explained: "Almost always it has been the case to have two or three possibilities that are put up for a student's consideration." Time is not wasted on topics that do not meet the research aims of the group, or on those that cannot be funded: "A range of available projects where the resources and equipment are available is offered, and then it is a matter of negotiation between the supervisor and the student" (an experimental physicist). The highly competitive nature of science research was well illustrated by a geology supervisor whose research was expensive and directed towards fast output: "In our department a student would never dream up a project. That is up to the supervisor, because the supervisor is driven by the funding he or she receives, and so consequently we can't afford to dally around with other whims. Basically, we would lose our funding." Far from being disgruntled with this convention, the students' views of this practice were strongly supportive: " . . . so I'm quite happy . . . the topic is my supervisor's . . . that's work that they're working on and I'm quite happy that it should be," and, " . . . the department received a large grant to set up new experiments. It was fairly frontier-type research, which was a very new field that people hadn't started in, so there was some potential. We were both (the supervisor and the student) in the thing together."

In stark contrast, the research enterprise in the humanities tends to be highly individualised. It requires independence of mind from experienced scholars, and the nature of the work means that it is often conducted in physical isolation from other people – for example, doing archival work in a library. Doctoral students are expected to exhibit the ability to do research under these conditions from the beginning. They are expected to come to candidature with a demonstrated capacity for research and to have advanced writing skills. From the beginning, they must also demonstrate the ability to work independently. An historian commented: "We expect them to work independently when we accept them. After all, if they are going to hit the wall at this stage, then there's no sense prolonging the whole thing." The concern that students should identify and develop their own topics also strongly reflects the expectation of intellectual independence, even in fields that cross disciplinary borders. A supervisor in archaeology, who identified his particular field as belonging to the humanities argued: "We believe that if you give a person a PhD topic, then you're inviting the worst trouble in the world. Students have to come up with a topic and that's the very first thing. They've got to believe that even though it might get modified a lot, it's originally their idea that they're working on. And this they find enormously difficult to do . . . I think it is the first measure of whether they are going to be a good student or not."

Supervisors from every discipline reported that topics are usually altered somewhat in the initial stage of candidature, though the extent of pruning or changing direction varies considerably. In externally funded science areas, there is little room for negotiation by the student. A geology student recalled: "I did my honours here and then the possibility of a PhD research project came up here which I thought was really interesting. So I chose to do that and then . . . while I was away overseas on holidays . . . my supervisors wanted to get industry support, so they put all this hard work in and actually got industry support and I was able to get an industry grant, so it's a bit more

complicated now." In areas where knowledge is not accretive and there is less competition for intellectual territory, there is more scope and more time for negotiating the
direction of the research. A discussion between two economics supervisors highlights
the greater scope and time that may be needed for topic determination:

... there is no preliminary statement of 'this is what the student will be working on'. And we ... will
really very seldom be able to make any confident prediction about what that student will be working on in
six months' time.

That's not always true. I've got a student started now and we've actually within two
weeks honed into a subject and actually defined an area to work in. I did have to
change it to a sociology. . . .

Greater autonomy is expected from humanities students, whose topics are
expected to be individualistic, particularistic and which may take some time to
evolve. A philosopher explained: "The student is the one who drives the thing. But
there is advice from members of the department; I think in the end the crunch is that
people often bite off just a bit much, which doesn't help them, so the judgement of
the supervisor should be respected. Philosophy being what is, by and large you can
decide to build great detail on some topic, and as you do that, the focus can change.
But that's all part of the thing."

An issue which arises from these differences and which has implications for both
supervisors and for students, is the extent of student commitment to the topic, particularly when there is little scope for choice. A biochemistry supervisor made the
point: "One of the things that science students have to come to grips with earlier . . .
relates to choosing the topic. They're intended to take ownership; women especially
don't take ownership of the project at an early stage . . . and we have to sometimes
figure out how to get them to do that, say 'this is yours'. I'll help you, but it's yours
and you will drive it."

1.3. Developing Research Skills

The approach in humanities fields may seem appropriately laissez-faire, but the
expectation is clear: the aim is for the student to make a firm commitment to the
research. The student is expected to take a position in relation to existing knowledge
in the field and to resolve this independently. One supervisor elaborated: " . . . they
have an idea during their honours degree or postgraduate degree, but I think it's much
more that's required of them at [the doctoral] level, and they have to get themselves to
that level largely on their own." The usual procedures adopted by supervisors for setting expectations for independent topic development and progress are also designed
to achieve the aim of discipline-appropriate behaviour. As perspectives are highly
individualised, students are left largely to their own initiative to determine, and seek to
have addressed, their own needs. A very experienced historian gave this account:

Well, when a new student comes to me and asks me to supervise the thesis, then
I make an appointment with them. We might go to lunch or something like that. We
spend a lot of time talking about how we each operate and how we will interact and
so on. I tell them they will have to work on their own, and how to check with me and

where they should go and that sort of thing. And I make it clear what they need to do to get started in a field and orient themselves in that field.

The practices of the humanities contrast strongly with those in science, where, in laboratory or group settings in particular, the aim is to ensure that the student can work within the parameters of the group and the field. This usually means that a prospective student is first assessed as being a reasonable 'fit' with the existing research culture of the group: ". . . we have a lab meeting after we've all talked to the prospective student, and (we) decide whether it's going to work or not" (philosophy supervisor). The student is given quite close direction in the first stages of the research because there is no time to be wasted. The rapid pace and competitive nature of the research enterprise in science is reflected in the practice of setting clear, often step-by-step, expectations for doctoral students to ensure that there is as little margin for error as possible, and emphasising the fast production of results. An experimental physicist reported: "The staff member has to be very responsible. The student has to be taught all the correct techniques. This will probably take the first eighteen months. You've got to watch them and look after them. You can't let them just try things out. You let it depend on how fast the student moves, how quickly they can stand on their own."

Students in science-based disciplines are rarely required to undertake coursework as part of their doctoral study program. They may be immersed from the beginning in the culture of teamwork which provides skills development 'on the post', or they may work with supervisors in a 'master and apprentice' situation, thereby also learning 'on the post' the skills they will need. The first priority in these settings is upon developing appropriate skills to undertake doctoral research. The emphasis given to laboratory-based coursework is exemplified in fields such as particle physics, where theoretical students are nonetheless required to learn experimental or laboratory techniques.

In contrast, it is usually the case in the social sciences that students must have acquired the appropriate methodological, technical and analytic skills, and the appropriate disciplinary knowledge, to undertake research in their proposed area. Because the range of specialisms is so diverse across social science disciplines, students frequently come to candidature needing to develop some background skills or knowledge. Becher (1993: 120) characterises this diversity in economics research settings in the United Kingdom:

> . . . the demand for common technical skills, especially those related to mathematics, appears to be on the increase. . . . a substantial training element was institutionalised in the form of a requirement for doctoral students, with a few exceptions, to complete a preliminary masters course before embarking actively on their own research.

Not all students are required to enrol in preliminary masters' courses, however. It is common to assess the student and ensure that appropriate avenues for skill development are offered; in some cases these are a requirement for candidature. Honours programs offer one mechanism by which students may acquire appropriate skills, and sometimes doctoral students are required to complete subjects at honours level or master's level. Even so, it is typical for the supervisor to assess the abilities of the student in relation to the proposed topic: "Now they have typically taken an undergraduate course within that same level, so they should have some idea what they are getting into. And if they don't, then we make sure they do some coursework topics from honours or masters, to get them to the point where they can move on" (a dean of business). An economist pointed out that: "Some students come with less skills that enable them to get on, and I insist that they learn standard skills. I say they've got to find out . . . and there's plenty of facilities for that to occur . . ."

Overall, very little coursework seems to be conducted for doctoral students in humanities disciplines, where research undertakings are idiosyncratic and scholars, like doctoral students, do not necessarily share theoretical or methodological frameworks. Becher (1993: 120) also found that the nature of doctoral training varied considerably across disciplines as a result of wide differences in students' training requirements and the need for a critical mass of students to make formal coursework programs viable: "There are seldom enough takers in any given specialism [in history, for example] to allow a close focus on particular training needs . . . nor is there usually a critical mass of doctoral students to make such programs viable."

2. FORMS OF SUPERVISION

In social science fields, knowledge claims are based upon methodological frameworks and theoretical positions with which the reader is expected to identify. It is therefore important to match doctoral students with supervisors with whom they are likely to share research interests and approaches. This accounts for sole supervision being marginally more common in social science settings, though there are many instances where there is a preference for joint supervision, and sometimes multiple supervisors, in the form of supervisory committees. Supervisors on the whole understand that it is not always possible to cover all the knowledge and methodological bases for students working in novel areas, a situation occurring especially in interdisciplinary disciplinary doctorates.

Comments representing political science and law are: "Expertise is specialised; one doesn't expect to be expert in everything . . . ;" and " . . . you look for appropriate supervisors, and if they aren't prepared to pitch in, then we have to reject the application." In applied fields, the establishment of supervisory panels or committees appears to be increasing, to ensure that a student's choice of theoretical and methodological platforms for the research can be properly covered. A business supervisor explained: "I was in the position of supervising a PhD whose ideological position was quite different to mine, so I ended my supervision. But often supervision is shared, and we get people in to advise about certain areas if we need to, and we've now got a committee . . . "

Joint and associate supervisions are also becoming commonplace in the applied sciences, where an experimental psychologist observed: "We often have students bounced off from other departments; we base our decisions (on sole and joint supervision) on expertise, just work out what we think we'll need. If a supervisor doesn't feel confident, we appoint an associate supervisor."

In science, though not in applied areas, sole responsibility for supervision is more usual. A physics supervisor, implying concerns about joint supervision, commented: "There are people I'd be less than happy to work with; we have different areas, that's how we are." General guidance is, however, likely to be a collaborative effort, and students are encouraged to take advantage of other sources of advice in the department or in the lab, reflecting the collaborative nature of much published research: "The people who've helped me most with the more mundane things are the support staff in the electronic and mechanical workshop" said one student. A soil science student explained: "I think you learn a lot from the other senior students."

In some laboratory settings, this process is systematic, as a biochemistry supervisor explained: " . . . we hope that we'll have a senior PhD student or a post-doc in the lab who can nurture new ones. That's a critical issue. If you don't have that senior layer in the laboratory it can be very, very difficult for the student." Sole supervision is qualitatively different, however, in theoretical areas of science, where students do not have team-based support. Sole supervision was characterised by a supervisor in mathematics: "We have the same world view, and . . . we see it very much as an apprenticeship. There's usually sole supervision."

In the humanities, the sole supervisor usually becomes the primary point of contact between the student and the research culture, and, in effect, the main source of guidance about the conventions of knowledge production in the field. Research is particularistic and tends to be fragmented within departments. The choice of topic constrains the choice of, and is constrained by the availability of, an appropriately well-informed scholar who is willing to make a commitment to the research. As a result, students may not be able to find more than one suitable supervisor. Furthermore, there are dangers in joint supervision because theoretical perspectives are so individualistic and the likelihood of clashes is high. A philosophy student recalled: " . . . we tended to have consultations together; they would sort of both be in the same room, and it wasn't working for anybody . . . I found that it was more productive if I saw them separately. Eventually . . . I had a bit of a dispute . . . and I had to choose." A supervisor in English literature commented on the importance of intellectual rapport, reinforcing the appointment of sole supervisors: " . . . the majority of students want to work in a new paradigm . . . there are some clear cases where the [doctoral] student should go to X or Y. You can only take on the student if you've got the primary ground. But also you have to keep an eye out for the emotional dimension." In the humanities, topic selection and finding an appropriate supervisor are frequently part of a joint process for students to negotiate. A cultural studies student explained: "I knew he was very enthusiastic about citizenship and he'd know a lot about East Timor. I kind of did my research about him before I went to see him. And then he didn't agree for a while – he was doing his own research on me!"

3. PROVISION OF RESOURCES

The characteristic way in which research is financed in a particular discipline has a direct influence on the funding of doctoral study in that discipline. A pointed comparison can be drawn between a student in experimental physics and a history student. The physics student enthused about conference attendance and the benefits of substantial financial support: "A year before my project started, the department received a large grant to set up new experiments . . . We were both [the supervisor and the student] in the thing together. I felt very excited about that . . . my equipment cost hundreds of thousands." In contrast, the history student lamented: " . . . there is a (spare) computer in the department but that was for staff only, but postgraduates do use it . . . which I didn't discover until months later. A lot of what you can have is kept quiet for economical reasons, definitely."

One very obvious difference in doctoral research across different knowledge areas is the degree to which the research attracts substantial external funding. In many countries, the formula for funding research overall is skewed markedly in favour of experimental science and medicine. Theoretical fields, whether in hard-pure fields such as physics, or soft-pure fields such as history or cultural studies, are funded as low-cost research areas. The sharp contrasts between theoretical, experimental and more applied research fields are, of course, extended to doctoral research. As described earlier, the nature of research in the humanities is highly personalised: the use of theory is individualistic, with scholarship in most situations being a lone enterprise. Sources of data such as archives and libraries are not usually particularly costly to access, unless fieldwork is involved, when compared with some areas of experimental science, such as geology, where equipment and climatic conditions present large costs. Research in the humanities has tended, on the whole, to be less well funded than in some other disciplines.

In areas where funding for research is limited, students are explicitly given to understand that resources are scarce. Students may find that they are immersed in a 'culture of poverty' as a result, but the norm seems to be widely accepted by them. One doctoral student reported: "Last year's new postgraduates came into the collo-quium and we just told them all these things, that they could get this and that, and most people were quite surprised. I mean, people who had been there two or three years didn't even know these things. It's definitely kept quiet." A history student reported: "There's a torrent of bad climate in the department . . . and it seems to me that any financial restraints that are having to be practised, are on students. It affects space in the department, access to photocopying, access to computers, those sorts of things. Attendance at conferences? In this department you don't get funded for that."

A noticeably less restrictive set of circumstances for financial support tends to pre-vail in the social sciences. The knowledge base is less certain than in science, so there is a heavy reliance upon theory and shared methodological frameworks. An important consequence is that experienced scholars tend to use scholarly networks to help them keep abreast of theoretical and methodological developments in their fields. Doctoral students too must keep in touch with new developments: in order to do so, they may need to be helped by taking courses or by arranging meetings with significant scholars. Typically, they are also expected at an early stage in their candidature to develop tech-nical and analytic skills that are versatile and enduring. With this in view, social science students are usually financially supported to some extent. A professor of economics commented on the need for shared skills: "We think it's an important area of their training . . . to equip them with transferable skills . . . (because) the economic histori-ans and the econometricians have got to be able to 'talk' to each other. . . . " A business supervisor reported: "Sometimes a small group have to go and do some kind of course, and of course we'd help them to do that. The *quid pro quo* is that they can probably share the skills around in the department once they've got them."

Conference attendance is also encouraged and supported if at all possible in social science settings, though levels of financial support vary across departments and from year to year. In some cases, supervisors reported that they did not make known to students the extent to which conference attendance, fieldwork or courses would be financially supported, in case funding sources dried up suddenly. A spokesperson for a group of social science supervisors in an Australian university commented: "Our students can apply for up to $700 from the school-allocated research funds, but that changes from year to year." A director of postgraduate research in business/tourism said, "We make them bid for a pool of funds for conference attendance, but we do put a $1000 cap on it". Even though resources are limited, there is a high value placed on skill development and on conference attendance, which to some extent overrides concerns about a lack of funds. A high priority is placed on being able to support students in keeping abreast of developments and linking up with scholarly networks.

In the sciences, there is a different set of financial opportunities in place to support candidature. As the nature of the knowledge base is more tightly structured than in other disciplines, researchers know the 'frontiers' of knowledge, and large research questions can be taken apart and tackled piece by piece. As a consequence, the race for new knowledge products in many fields is fast and competitive, and scholars must be quick to make new knowledge claims. In experimental areas, fieldwork and equipment may be expensive, and, though funding tends to be more accessible than in other disciplinary areas, research cannot proceed unless the availability of adequate funds is certain. These characteristics are duly reflected in the budgets available for doctoral research. A physics student expressed a key disciplinary value when he explained: "You know everybody has a budget, we have to live with that. I'm afraid one of the important functions of experimentalists is to keep them[selves] well funded."

In experimental science fields, doctoral research projects can only be approved if external funding is guaranteed. Students soon come to accept that this is part of the research culture in which they have to operate. A soil science student commented: "A project will fly only if it gets external funding, and that is a very stiff process because . . . a staff member might have an idea that might involve a higher degree student . . . but, if the money doesn't come, well then the project dies." Though some science departments are more generous than others, there is usually a typical procedure in place to ensure that students know what they can expect in terms of financial support, as well as what is required of them in return. A geology supervisor stated: "We expect students to give poster presentations and we try to get them off to at least one national conference during their time. . . . "

There is a wide range of other resources made available in support of doctoral research, such as equipment or financial support for fieldwork, the allocation of funds in support of student meetings and research presentations, stationery, typing of transcripts or technical assistance. While norms vary according to the nature of the field, the overwhelming target of resource allocation is to initiate doctoral students into the conventions of knowledge production in the core discipline. Financial support, for example, tends to be conditional upon the presentation of papers and posters at conferences, even though they might be small, local meetings, because the expectation is that scholars will keep abreast of new developments in the field and build helpful scholarly networks.

In the more autonomous settings of humanities research, there is often a paucity of funds, leading to limited financial support for doctoral research. This is to some degree offset by the fact that the more slow-moving nature of research does not often require the constant maintenance of scholarly contacts, nor does it call for expensive equipment. There is a marked difference in available resources in support of student research endeavours: the limited resources usually available to humanities departments have resulted in situations where humanities students have not even been able to be provided with a desk in their department.

Humanities students frequently report that the extent to which they feel supported in their endeavours depends heavily upon their supervisors' own efforts. These students tend to rely on their supervisors to provide a conduit to the specialism and to its conventions governing the production of knowledge. The style of supervision in these cases is necessarily closer and more personal than in the sciences. In one history department, funds were so limited that it was standard practice to write letters of introduction on behalf of doctoral students to ensure that they could make contact with significant scholars in their fields. Supervisors in humanities settings frequently reported taking personal responsibility for ensuring that their students were given some measure of financial backing, and were supported by a departmental commitment to their work. A supervisor in philosophy reported: "We've tended to be close to our students . . . We try to get them teaching experience. It also helps with morale, and we support them with an allocation of research funds . . . it's based on the merit of the project, and they compete with staff."

The nature of knowledge production is clearly related to the funding of doctoral research in theoretical areas of science. Sharing many characteristics with knowledge production in the humanities, where research is a highly individualised and, therefore, a more isolated endeavour, doctoral students are given to understand the implications of the limited funding available. A theoretical physics student reported: "We just don't get the opportunities [of our experimental counterparts] . . . sometimes somebody internationally renowned comes along, but that's it . . . You often feel that you see your supervisor and that's it." In these settings, students are heavily dependent upon their supervisors for intellectual resources and guidance. Where funds to support research are limited, supervisors report taking greater responsibility for making sure that the student is on a sound footing and can handle the level of independence required.

Knowledge production in highly specialised research fields requires quite specific resources and supports, and, hence, the norms are appropriate to the nature of the knowledge base. In experimental science, funds are a primary consideration in the allocation of supervisors and in determining topics, and supervisory roles are largely confined to ensuring that the necessary pace of progress is kept to maintain deadlines and funding requirements. Because much of the research is collectively produced, doctoral students are supported by significant others, such as post-doctoral students and other staff, in the research setting. The role of the supervisor is directive until the student is able to take control of the project; the values and norms of the research enterprise are made explicit and are closely monitored.

4. HUMAN RESOURCES

It is clear that students accept the prevailing conditions for research in their fields, even when those conditions do not seem to match their own priorities. Doctoral students from the beginning accept the ethos of science, rapidly adopting key values and norms that are explicitly part of the research setting and clearly accepted by others in the research environment. In these circumstances, various individuals in the research environment, including the supervisor, provide opportunities for learning the rules of the game. Becher, Henkel and Kogan (1994: 74) further elaborate this idea:

It is clear that research education in physics and biochemistry is deeply embedded in the organisation of research itself. Students are incorporated into research groups, some at or near the frontiers of the subject, some within fields in a state of rapid change, some, in physics, in applied research. The dominant group-bases apprenticeship model reinforces a pattern of early specialisation. . . . Disciplinary practice tends to support Clark's suggestion . . . that it is the requirements for knowledge production rather than education that frame the doctoral student's experience.

In the humanities and in areas of theoretical science, research tends to be less expensive overall, being more esoteric and highly individualised. Ideas may be slow-forming and the autonomous nature of research requires strong commitment to the doctoral topic and to its completion, as well as intellectual and emotional support from supervisors. Because many doctoral students work for long periods in isolation, the role of the supervisor tends to be more directed towards providing encouragement, intellectual validation and appropriate background to knowledge in the field. Becher, Henkel and Kogan (1994: 80–1) describe how these characteristics operate in the humanities, noting the inherent isolation in doing history, describing it, along with modern languages, as individualistic, having little need or potential for research organisation and being a relatively low-cost enterprise. This is indeed the situation in which doctoral students find themselves, and, though they may find the setting lacking in resources and limited in the range of intellectual support, these explicit norms are soon accepted, subsumed by a strong commitment to individually determined research topics. In these circumstances, the supervisors' role in supporting the research is considered particularly important by students.

In the social sciences, resources must be invested to ensure that doctoral students have opportunities to develop appropriate methodological and theoretical platforms for advancing knowledge in specialised fields. Because the fields themselves are so diverse and so dynamic, it is necessary for students to keep abreast of developments as they advance their own work. The key role of supervisors is, therefore, to facilitate the development of appropriate technical, analytic or methodological expertise, and to ensure that the students know about recent theoretical developments, often through making contact with other scholars in the field. In some cases, particularly where students are working in novel or transdisciplinary fields, joint supervision and supervisory panels or committees are established to cover, as far as is practicable, the range of expertise required. These forms of co-supervision also reduce the extent of responsibility expected of a single supervisor who may not have much experience or expertise in the new field. In other cases, though, it is simply not possible to cover all the ground. There is little evidence of collaborative research in social science fields, such as economics and sociology, for example, which might offset a lack of supervisor competence.

The question arises about how doctoral topics in novel or transdisciplinary fields can be effectively supervised in the light of the global trends towards interdisciplinarity and applied research. The responses of different higher education systems differ. In the United Kingdom, effective doctoral training is thought to be assured through recently imposed, mandatory coursework components of the degree. In Australia, policy directives have concentrated funded research in a limited number of key research areas. Evidence about the merit of these developments is yet to come in, though such organizations of the research endeavour have the potential to provide well-defined research cultures for doctoral study.

5. PATTERNS OF RELATIONSHIPS

In almost all cases, supervisors reported that it was important to develop close, but not necessarily personal, relationships with their students as a means by which the supervisory role could be more easily maintained. Upon closer questioning, however, the characteristic roles of supervisors were clearly evident: they facilitate different needs for the process of knowledge production. In laboratory-science departments, meetings with supervisors may be frequent and fairly informal, with students being encouraged to learn from and be responsible to their research groups. A science student described this process: "The three of us are in the same research group. We have meetings every week and if there are any questions about how to do something, or what your results mean, and you want to discuss them or anything like that, we do that every week." In most fields, there exists a hierarchy of scholars, from those with exalted research records to the newly initiated doctoral students. A chemistry supervisor illustrated these values when he claimed: "It's important that relationships [with students] are close. We have an avuncular relationship with them. By the end of their PhD, they call us by our first names . . . there's a rigid hierarchy in this field. It's well defined."

In the social sciences, however, students are just as likely to be working in isolation as in a small research group, and the role of supervisors is to ensure that students are given appropriate opportunities and resources to make progress with their research. A supervisor in business commented: "It's not a spoken thing, but we (the supervisors) all know it. . . . You join as the office boy, you get a lot of on the job training . . . courses, one-to-one, and students sometimes teach each other as well . . . so there's almost a market solution here. . . . " An economist described the patterns of supervisor/student relationships: "Sometimes they can get to be friends and sometimes they might not get to be friends. It just varies. It depends on the individuals and whether they get on together." The important element of supervision for the social science supervisors was to ensure that their students had adequate intellectual resources to meet the expectations of their fields.

In the humanities, developing a rapport with the student is a primary objective for supervisors because of the heavy reliance on the supervisor to ensure that norms are clear and that progress is being made in a field where there may be few other active scholars. A philosophy supervisor described this: "We tend to have fairly close relationships with students, some are friends and are quite close, but some are friendly but formal. Research in philosophy is very personal, so what you want to know is if there's something showing up in the work that you need to sort out. You have to have a relationship for that."

6. REGULATORY CONSTRAINTS

There are important procedures in place to ensure that doctoral research meets certain norms and standards, though these differ according to the research setting. Mechanisms for ensuring that theses are of acceptable quality are institutionalised in the form of written or implicit procedures that supervisors can invoke when necessary.

In most science departments, it is not always possible, especially if the students come from another university, to assess accurately their preparedness for doctoral study. One mechanism, prevalent in areas such as agriculture, biochemistry, geology and microbiology, which are heavily dependent upon external funding, is to enrol students in masters programs, with conditional offers of upgrades being made to the more able students after one year. This practice is also prevalent in areas where students come to doctoral programs from fields of professional practice or from industry – areas such as in education, economics, business and legal studies. In some social science areas, as noted earlier, students may be required to demonstrate adequate technological and methodological skill by enrolling at master's level and upgrading once some of the chapters of the thesis are written and the standard judged appropriate. In other cases, students are conditionally enrolled with the requirement that they attend or obtain credits in certain courses relating to analytic, technical or methodological competence. Another mechanism, which typically occurs in theoretical areas of chemistry, mathematics and physics, but is largely confined to the sciences, is to recommend downgrading to a masters degree if the student's work has not met acceptable standards. One chemistry supervisor commented: "Students are advised to write up an MSc. It happens occasionally. In chemistry, an MSc may be a failed PhD."

In the humanities disciplines, upgrading from masters to doctoral programs occurs, but is less common. The doctorate has relatively recently eclipsed the masters degree, as Simpson (1983) shows; students who are not sure of their ability or aspirations are more likely to begin at masters level. Those humanities students who consistently do not make sufficient progress in the doctoral degree are often given extra advice and assistance by supervisors, to the point of allowing candidature to run out. There are few terminations of candidature, and it is felt that allowing students themselves to "fade away", as one supervisor put it, is far better than making failure public. Supervisors from various fields explained this practice: ". . . we try to resolve what we can . . . we would tell them [the students] they are not performing, and of course, avoid confrontation." and "Once we had a termination. We bent over backwards . . . appointed a co-supervisor to help with the writing part of it . . . but we were in despair. It resolved ultimately with the student withdrawing, which was a relief to everyone."

7. FIELD-SPECIFIC INDUCTION

From experiencing how topics are normally identified and developed to learning from supervisors and others, students are immersed in research cultures, each with its own traditions and conventions for producing and communicating knowledge. Through the opportunities formally presented to them during candidature, such as working collaboratively, as in experimental science, or autonomously, as in history, students are able to identify with particular disciplinary values that are central to the nature of the research enterprise. They are also able to learn important skills and capacities that are considered to be identifying features of disciplinary expertise. In some cases, supervisors take responsibility for supporting these learning opportunities; in other cases, the opportunities are so much part of the research culture that students are simply expected to participate in them. Nevertheless, there are recognisable pathways through which doctoral students are inducted to the disciplinary culture of their fields, and supervisors appreciate the importance of these pathways. Other pathways, less explicitly part of doctoral programs but nonetheless a critical part of socialisation to a particular discipline, are also an important aspect of being immersed in a particular research culture. These pathways, and the values they embody, are addressed in the following two chapters.

CHAPTER 5

EXPLORING THE RESEARCH ENVIRONMENT

1. SOCIALLY-BASED LEARNING

For the student trying to get to grips with the rules of doctoral study, there is much to learn. Alongside the need to master and extend a particular knowledge territory, there is also the need to know how to behave appropriately in a knowledge community from which the student wants to earn acceptance and approval. There are, therefore, critical lessons to be learned about how to fit in to the new environment and how to profit from what that environment offers.

Every institution has a particular character and every department has its own direction, systems and personalities. Within this framework, students must learn the common traditions, conventions, codes of professional conduct and objectives that are central to the broader international arena of their particular specialism. Familiarity with the relevant rules of knowledge production is essential to being able to produce an acceptable doctoral thesis. Students must not only conform to these rules in order to gain acceptance of their work, but they must also develop a clear understanding of the boundaries of what is acceptable. To appreciate the magnitude of the learning that takes place, and to identify the kinds of opportunities that enable it to happen, it is first necessary to consider some core disciplinary characteristics in field-specific settings, so that we can better recognise how students come to find them out.

We have seen earlier how a framework derived from Bucher and Strauss (1960) helps us to understand that doctoral students are inducted to the complex values, norms and conventions of academic specialisms. Bucher and Strauss (1960:326) account for the "many identities, many values, many interests" of academics in disciplines who at the same time lay claim to unity. Becher's (1987a, 1989a, 1990b) portrayal of disciplinary cultures identifies some core characteristics of disciplines that are at the same time dynamic and subject to sometimes-rapid change and development. The implications of understanding these complexities are significant in that they speak to the level of sophistication of learning at the doctoral level. They also point to an emergent feature of doctoral research in universities today, namely, the increasingly applied or transdisciplinary nature of research topics.

Since the rules for knowledge production are highly field-specific and appropriate to the social and communication patterns of scholars in specialised fields, the framework of this enquiry includes both the processes which socialise individuals to the core activities of knowledge communities and those which give individuals a clear understanding of the boundaries of accepted behaviour. Both sets of processes relate to the cohesive, core elements of knowledge communities such as disciplinary identity, values, codes and conventions, as well as to elements that may be at the boundaries of acceptable conduct, such as unconventional methods and novel techniques. An investigation of how these processes of socialisation are assimilated by students in the absence of explicit references to them in formal aspects of doctoral programs or in supervisory meetings helps to explain the nature and value of the tacit learning undertaken by doctoral students.

1.1. Initiation to the Culture

The social setting of science, in so far as it involves laboratory research or other teamwork, offers many advantages for students being initiated into the values, codes and conventions of a scientific field. In particular, it provides opportunities for skill development and for learning appropriate techniques and methods. The first encounter with teamwork in science is highly illuminating for students, several of whom acknowledged that they were assessed by their supervisors, and introduced to the team they hoped to join, with a view to identifying their 'neatness of fit'. The emphasis upon producing results and meeting deadlines means that neither is there the time nor are there the resources for taking on students who might not be productive and who might not be able to accept the priorities of the team effort from the start. A geology student reported: ". . . and the minute they got on to me, I came over, met with them [the team] and had a discussion about what I want to do, what they do and how they do things . . . because they are people who get on together. They are people who can help each other. Actually, there was no acceptance at that first meeting, but later it got sorted out."

Because there is no such social setting for research in theoretical areas of science, supervisors frequently reported having discussions at or before enrolment to establish the conditions for research. One supervisor in physics said: "The supervisor is the one who is going to be stuck with the student, so the supervisor has to explain what it means to be an apprentice. We expect the process [of doing research in the department] to be well understood."

One feature of scientific research made explicit to prospective students at this stage is the hierarchical nature of scientific knowledge communities. Doctoral student status in these hierarchies is made abundantly clear. A zoology supervisor commented: "Basically it's an apprenticeship. I doubt if things are negotiated. They are encouraged to talk to other students. . . . By their third year they might call us by our first names. It just depends." Other science settings may be less hierarchical, but the status of the student in decision-making is abundantly clear. A fisheries supervisor reported: "You can't have the students doing their own thing at all. We work to contracts, to deadlines and to (funding) grants. The project leader makes the decisions about intellectual property, work schedules and so on. The research group works within that. And when a PhD student joins up, they know they're coming in to learn, so they do what they're told.

The prevailing values in the social sciences are distinctly different. The eclectic nature of social science research described by Becher, Henkel and Kogan (1994: 76–7) requires that the setting for doctoral study is one in which an individual must be well armed methodologically, technically and with a sound knowledge of theoretical developments before embarking on doctoral work. Supervisors are generally open with students about the difficulties of research in the social sciences: as an economics supervisor depicted, "Economics is a relatively new discipline . . . it has to be seen to be more theory based than, for example, philosophy, and we're very tough on what we expect [from doctoral students]. There's all these sorts of very, very critical, very tough barriers around that we occupy academically, and hence . . . students have got to be survivors."

The nature of supervisory relationships tends to be less rigidly defined than those in science, reflecting the less rigid status distinctions and illustrating a more avuncular attitude towards doctoral students. Supervisors were committed but patronizing. A supervisor in management commented: "I think we are very caring of our students . . . we have a reasonable internal culture, we put ourselves out for them, try to get them teaching hours, treat them like junior colleagues and help them to learn appropriate things." Connecting students with developments in their fields is also important: "I'll often fill in the background . . . It's obvious to tell them about relevant societies and meetings, conferences and so on." (an economics supervisor). Supervisors recognise that the diversity of research endeavours in the social sciences is not by itself conducive to an atmosphere in which students learn from each other, though learning from each other is a feature that many encourage, both for expedience, and to counter the dangers of intellectual isolation. Opportunities for social development within departments are often created to provide something of a research community in which students can learn from each other's efforts.

The overwhelming impression in the humanities was that doctoral students were regarded more as new colleagues by supervisors who attempted to establish a positive rapport with them. Supervisors reported emphasising how the research might be resourced and how the student might be situated intellectually within the department. This concurs with Becher, Henkel and Kogan's (1994: 78, 81) view that, at least for historians, there is a strong sense of disciplinary unity and solidarity among scholars, for all their individualism. In the initial meetings with supervisors, students tend to be inducted to an ethos in which they feel valued. A history supervisor explained: "The first thing I do [when a new candidate arrives] is to sit down with them and have a talk. Find out where they are up to. What they're interested in. What they've done. What their life is like. Things like that. Then we might have a talk about what I am doing, where the field seems to be at the moment, and so on. And we just talk generally about doing history. . . . That's important because they have to have a sense of . . . people doing history. And then there's a lot to build on later on, because lots of things happen during the course of a PhD."

The importance of the departmental research environment in revealing key disciplinary norms was illustrated by a professor of biochemistry who could compare the performance of students in her specialism in the department with those in an associated research institute. She explained: "I wouldn't say it's a stable situation, but having students working within the department and part of the local culture . . . [makes it] easier for them to support each other . . . it provides a good conduit of information. . . . Students at [an outside research institute] are under pressure to produce products. The values are different . . . there is just not the culture there of looking after the PhD student."

These differentiations in the settings for doctoral research are consistent with those reported by other researchers. Becher, Henkel and Kogan (1994), for example found research education in the sciences to be closely enmeshed in the overall organisation of research. In contrast, in the humanities and social sciences, the individualistic nature of the enterprise means that graduate research is usually distinct from that of established scholars. These researchers also argue that in consequence, doctoral students in the latter fields have greater expectations of their supervisors. This is clearly reflected in the attitudes that supervisors adopt towards their students from the beginning of candidature.

1.2. Norms In Work Activities

By immersing doctoral students in the process of knowledge production, supervisors also immerse them in the dominant values and attitudes of specific disciplinary fields. Becher (1989a: 25) argued that disciplinary ideology and cultural capital are essential to effective socialisation. Students' comments about what they have to learn and which behaviours they have to acquire give an insight into how this works in practice. Gerholm (1990) adds that an understanding of disciplinary ideology and cultural capital is built principally through the development of tacit knowledge of the technical aspects of the discipline, and of the personal aspects of how to behave in the formal and informal settings of disciplinary work. Students develop this kind of knowledge in many ways. Social science and humanities students, for example, have to know how to handle conflicting rules; they have to participate positively in research seminar discussions, even though collaboration in their fields is subtly discouraged.

Because it is fundamental to the notion of disciplinary culture, becoming savvy in a particular field is a fundamental element of learning during doctoral study. The importance of tacit learning cannot be overestimated, because it is in this way that doctoral students eventually acquire acceptable styles of formal and informal communication and assume an active role in contributing to disciplinary knowledge. We explore how they do this in more detail in Chapter 8, but at this point the key question is, "how do supervisors ensure that their students become savvy"? Students' and supervisors' own descriptions of what happens in the setting for doctoral research give much food for thought.

In the humanities, a close intellectual relationship with the supervisor is the vital ingredient in developing a sound knowledge of developments in the field, and in understanding how to situate one's own research in relation to existing work. One supervisor explained the typical process in history: "We may well make comments in a letter . . . about how the focus might be developed and we would be aware of [what] the student needs in pursuing it, or who else they might need to talk to . . . they mainly need to develop a background." The priority is for supervisors to feel confident that doctoral students have enough resources to work independently, as established scholars do. Students typically comment that doing research on one's own can be a very lonely pursuit: "It was that feeling of intense and enclosed activities, . . . it was very isolating. I'd spend ages sitting in a library and come out and heavens, it's dark, you know." (a history student). Students apparently expect to work autonomously and in relative solitude. Their concerns are not so much about the intellectual insularity, which they view as part of the usual research process in the humanities, as about the isolating effect of not being connected with the research communities in their departments. One student reported: "I thought that by doing (my doctorate in) critical pedagogy would mean that I would fit right into the research culture upstairs. But now I find I go into my own bubble and don't come out for eight or nine hours. It feels peculiar to drive home and feel like I am driving back into my real life!" This effect is sometimes reinforced in practical ways by the lack of basic resources such as office space and computers, further attenuating the links to their research settings.

Humanities students widely report that having a positive rapport with their supervisors greatly compensates for the sense of intellectual isolation. A history student reflected on her experience with her supervisor: "her strength was . . . in the criticism of what I was writing, and . . . she would just support me. . . . This isn't very clear, but communicating, that was her strength." The emphasis in humanities research generally is upon developing a deeply personal understanding, which, as Bazerman (1981) describes it, evokes persuasion and insight.

The emphasis in supervision is very much the same, in that the support and commitment of an experienced scholar provides a critical touchstone against disciplinary expectations. As earlier noted, knowledge in the humanities is organic, holistic and tends to be constructed according to a vision 'from the top down'. Doctoral students have to determine what constitutes an acceptable 'vision' in a field, and it is largely through the commitment and guidance of an experienced scholar, usually the supervisor, that the unspoken boundaries and rules of acceptability can be learned and confidence developed. A good example of this was provided by a professor supervising a doctoral topic on Martin Luther King Jr: "I can think of folk who I would never get to examine a thesis on King, for example, or on an aspect of King's work, because fundamentally they dislike King so profoundly and their scholarship is coming from a completely different value set and historical background. These can be people who are politically either on the left or on the right, and because our discipline is in so many ways individualised, this can happen. So you work with the student and in the process the student learns their way around the movements and builds a path of their own, and knows the alignments, you see."

In the research settings of social science, the typical work activities of scholars emphasise different aspects of the knowledge base. The strong dependence upon theoretical structures and methodological frameworks results in a need to advance appropriate technical and theoretical bases for research. To do this confidently, scholars must remain abreast of current developments in dynamic and changing fields. Doctoral students must learn to work within these conditions and to identify a theoretical perspective that is appropriate and justifiable. The challenge is described by a supervisor in econometrics: "Even when you think you're shooting perfectly at the bullseye, you're relying on other people to actually keep you informed, guide you and tell you whether you've hit the bullseye. And once you've put it [the thesis] in . . . you can't wake up one night two days after you've submitted the thesis and say, 'I forgot to mention all that stuff, or rewrite the program to the new development'. . . . So all of those things make it very nerve racking; it's surprising anyone ever has any self-confidence. So we have to help them to become skilled craftsmen who can do different things, and we send them to conferences as well. They are two things that I guess we actively do."

In the main, students generate independent research interests, though they have to draw upon the skills and expertise of the immediate research community and the wider scholarly networks to develop confidence about the appropriateness of their work: practices which closely parallel those of established researchers in the social sciences. Research seminars, colloquia and conferences provide scholars with avenues to address these needs, and in most settings supervisors encourage students to participate in them. The eclectic nature of social science research often gives rise to cliques within departments, as well as within the parent discipline as a whole, and doctoral students are soon faced with having to deal with and learn from encounters with these cliques. Being very much the junior colleagues in the research community, they generally need to take the initiative and seize the opportunities around them, learning new skills where they can and making useful contacts where possible. While supervisors offer background support, students are expected to behave like experienced researchers and to be 'self starters'.

The setting in the sciences is different again. Though students are plunged into the culture of deadlines, rapid production of results and the need to mark out intellectual territory, they are also unequivocally treated as new and raw recruits in the research enterprise, at least for the first year or so. As described earlier, research is frequently a collaborative effort, which ensures that production values and outcomes are maintained. Doctoral students not only comply with the prevailing conditions for research, but, over time, they also demonstrate that they do not need direction and are capable of independent outputs. There is no expectation of independence at first. A psychologist with a strong experimental science background described the process: "There are ways of doing things. [Doctoral] students have to learn techniques, laboratory procedures, and more generally, ways of doing things. It's not just chemical or physical, they just have to learn how to manage. If you haven't gone through this indoctrination system you think you know it all, but you don't."

Part of the 'indoctrination system' is the process whereby a doctoral student learns to take a junior role in the existing research community. Even in theoretical areas of science, where doctoral students are more systematically mentored than in laboratory and team settings, they must work within the established hierarchy of scholars and meet their expectations. In experimental areas, the normal work expectations of turning up for an eight-hour day each week, of accepting responsibility for producing publications collaboratively or individually, of producing results and of meeting deadlines are also imposed upon doctoral students. As a result, it is not uncommon for students to report after a year of candidature that they are the "unpaid labour" in their departments.

In laboratory or team settings, students also have to participate in such cooperative activities as attracting research grants, producing results, and engaging in the information and skill exchange of their research community, fitting in accordingly with its production-line ethos. While doctoral students in the sciences begin at the bottom of the ladder, they are nonetheless surrounded by opportunities to master important disciplinary competencies, such as technical skills, grantsmanship and collaboration, and to observe the conduct of successful scholars. In general, they work in an environment where the pace of research dictates that they will not be permitted to stray too far from the norm.

2. PATTERNS OF COMMUNICATION

The literature on academic disciplines (see for example, Whitley, 1984; Becher, 1987a, 1989a) provides ample evidence that the predominant types of scholarly publications vary from discipline to discipline. Biglan (1973b: 211) observed how different forms of publication, styles of explanation and rates of productivity apply to different disciplinary areas, while Becher (1989a) explained why different forms of written communication are valued in different kinds of knowledge communities. Articles are predominantly the domain of specialisms where there is a dense concentration of scholars who work at a fast, competitive pace, with a "high level of collective activity, close competition for space and resources, and a rapid and heavily used information network." (Becher, 1989a: 78) Conversely, books tend to be more highly valued in specialisms having relatively few scholars in any area, a wide choice of research topics and little competition – and hence where the pace of research is slower, and " . . . researchers are liable to become engaged on long-range issues which may take years to puzzle out." (Becher, 1989a: 79).

While the nature of knowledge production and its influence on the patterns of communication in their specialised area may not be immediately obvious to doctoral students, it is necessary that they subscribe to the dominant values of their fields from the outset. In science, the pace and competitiveness of research and the relative narrowness of its frontier require that journal articles are produced to mark out intellectual territory, even during candidature. In the humanities, where the range of research topics is limitless, competition hardly exists, and research can be a slower, more deliberate and lengthy process. These values are reflected in the attitudes assumed by doctoral students towards publishing during candidature.

Whitley's (1984) characterisation of scientific fields – as organisations that delegate considerable autonomy to individual scientists for establishing and maintaining their reputations – provides a basis for understanding the codified nature of authorship in science. A key feature, exemplified by laboratory science, is collaborative publication. Collaborative authorship is a mechanism by which students actively contribute to the knowledge products of the team, in accordance with the objectives of science: to produce and report results as quickly as possible, and to claim intellectual territory.

Thrust into a culture of 'publish or perish', students soon find themselves obliged to co-author papers, at first to meet the expectations of supervisors or of team members. It is not usually too difficult to satisfy these requirements, because supervisors and more experienced team members are willing to lay down the rules on how to go about the process. An experimental psychology student reported: "I have research supervisors who have the attitude that the first objective of the PhD is tied up with the research strategy. You should know how many publications you're going to get out of the thesis, etc. . . . so it depends what you're aiming for." A physics student stated: "Oh, in physics always you have to publish. Of the student group, you all sit down, have a chat and they're always willing to help you. That's how we do it." A geology student crystallised the importance of publishing, for both students and supervisors: "a person [a supervisor] has what is maybe a three- or four-year contract and at the end of that their research record is put on display . . . so it is to [the supervisor's] own advantage to try and get as many papers out as possible. It's also a matter of funding. If you can show your students have been writing up their work, you might get the funding you need."

In the social sciences, there is less commitment to publishing during candidature, though supervisors agree that it can be highly beneficial to the student – especially so when the student aspires to academe. Specialisms where collaborative research and joint authorship are more prevalent offer exceptions to the norm of not publishing during candidature. Theoretical and quantitative economics generally are fields where not only are publications encouraged, but it is also possible for students to submit for their doctorates collections of a small number of publications on a topic.

In social science, in general, joint authorship is dependent upon the nature of the field and the extent to which the knowledge base and theoretical perspectives are shared by distinct groups of scholars. In some settings, the conception is mainly of research as an individual pursuit, even though there may be shared theoretical frameworks. This conception is reflected in the patterns of authorship, where, unless the student is exceptionally able, publishing during candidature is seen as a diversion from the central activity of completing the thesis. In principle, the policy is supported, but in practice students are only urged to pursue the idea if a chapter turns out to look like a publishable article.

By contrast, in other knowledge areas, the competitive nature of the field calls for publication in the course of the doctoral program. Giving papers at social science seminars and conferences is encouraged, reflecting the typical patterns of communication among networks of scholars. Experience of this kind provides valuable feedback to new scholars, but there is also a need to mark out intellectual territory. One social science student reported: " . . . unofficially, I think if anybody shows interest in doing discussion papers and trying to get things published, that's encouraged. But I'm starting to see now that I've only been doing conference papers and discussion papers." Another student commented on the importance of gaining recognition for new work: "What it means is that when I go to conferences, people have all got labels on them. Now, . . . if you do something different next time, suddenly you're more of a person to everybody at the conference."

In social science settings, less formal modes of interchanging ideas among scholars are an important addition to communication through refereed articles. Doctoral students must accordingly learn to weigh up the benefits to themselves of participating in the various alternative forms of scholarly communication, especially considering that publishing during candidature is more highly valued in some specialisms than in others. Where there is less expectation of publishing to mark out intellectual territory, time invested in developing and presenting work-in-progress can provide the feedback, advice and fresh ideas from established scholars that are essential to learning important disciplinary skills and customs. An education student explained, "If you don't publish, you don't get to increase your [curriculum vitae], but (if you do), you do get lots of good ideas and you find out how things are done. You get a lot of leads. You sort of get accepted socially as well as intellectually."

Where publishing during candidature is generally the expectation, not only are students able to mark out intellectual territory, but they also learn the importance of keeping at the forefront of technical, analytical and knowledge developments. An economics student described the process of keeping up to date: " . . . people publishing in journals who seem to have the ability to pick what is the flavour of the month in terms of those journals. You need to get to be good at that because . . . if you don't produce what in terms of the current thinking is acceptable, then you don't get published."

In the humanities, the degree of individualism in research means that there is rarely a critical mass of scholars with shared research interests within departments. Supervisors well know the difficulties of maintaining in-house departmental colloquia and seminar series, and take the view that an effort must be made to maintain contacts on the broader disciplinary level. Becher (1989b: 275), consistent with this claim, observes that communication among historians is largely through monographs and inter-institutional colloquia. The relevance of the latter is evident for doctoral students, given that their need for affirmation is understandably greater than that of experienced scholars.

Humanities students share concerns about being intellectually isolated, about needing feedback on their doctoral work and needing to feel connected with the research community. As one remarked: "Sometimes, if it wasn't for the other students, I wouldn't have continued." Another said: "There aren't scholars [at colloquia] in my area, but in a lot of other areas . . . it helps me because sometimes they'll be talking about the system and how to write as researchers." Humanities students have to take the initiative, seize opportunities, and to some extent create them for themselves, in a research culture that is particularistic, and where the professional societies are likely to be relatively weak and widely dispersed. Like their more experienced scholarly counterparts, students can seldom discuss their own research in depth with other students or scholars: however, being able to talk generally about their research provides them with some affirmation of their doctoral work. It is not difficult to see how students fit easily into the scenario described by Becher (1984: 189) of communication among historians:

Lacking the specialised knowledge of detail which could enable them adequately to discuss a colleague's research, they can share experience only at the general level of exchanging notes about sources, library collections and the like.

3. MODES OF AUTHORSHIP

Becher (1989a: 99) provides an overview of the norms of collaboration or sole authorship in different specialisms and accounts for important variations in the patterns of collective authorship of scientific papers. While many university scholars may not be consciously aware of these patterns, doctoral students come to know them well. Students are certainly knowledgeable about codification within their fields, and most can explain the significance of the sequencing of authorship in the relevant research papers. In science specialisms, where collaborative effort is most usual, there are curious variations. In some cases, it is a matter of citing as first author whoever has made the greatest contribution: but different, less obvious patterns are the norm in other fields. For example, a geology student explained: "It doesn't matter how many authors there are, really. The [doctoral] student comes second last. We all know that. It tells people who the PhDs are, and if they're any good." A physics student stated: " . . . most supervisors have a policy of putting their name last because in our scientific community that is the code for work of the supervisor". The pattern is different in chemistry, where a professor stated: "I grew up to think that alphabetical is best, it stops any arguments, but I know the American practice is generally to put the senior author's name last."

In settings where collaborative research and authorship take place, supervisors agree that it is important to acknowledge students who are promising by publishing with them, which of course has the added advantage of advancing their own reputations. Doctoral students are well aware of the benefits for themselves in this arrangement, in that much is learned in the process. A genetics student explained: "(my supervisor) has got a CV as long as your arm. But our group does all the work and he just gets his name on the paper automatically. Not that he doesn't give us ideas, but we do all the work . . . But then, it is good for us, too, because he's got a DSc and he's a star. And people see that you're publishing with a star." It is clear that an important entree to networks of scholars is by reputation derived from publishing.

Publishing with senior colleagues is a means of achieving credibility. A physics student explained: ". . . there has to be a balance. You should never feel cheated that somebody is cheating on the names, because usually you get more out of them than they get out of you. It's about reputation. But at the same time, they don't need it, you do."

Sole authorship during candidature is less common but can bring special reputational benefits. A physics student explained: "Recently, we met with this group in Germany . . . and came back with a couple of ideas and computer programs and things like that. So you're starting to make connections. And I found that my papers that I wrote were a reference in other places. So it's just being in the right place at the right time. But if what you're doing is a dead end, then no amount of posturing is going to make it interesting to other people."

It is especially important in theoretical areas of science and mathematics, where collections of already published papers frequently form the doctoral thesis. It is the norm here to obtain appraisal from distinguished scholars in the field. Not only does it provide a means of improving the component parts of the thesis, but it also helps to ensure that the student, whose intellectual territory is already marked out, may be well regarded in the discipline by the time the doctoral thesis is submitted. A professor of mathematics underlined the point: "We've got standards, full stop. And they're very old standards. That's why I encourage students to publish beforehand [before submitting the doctoral thesis]. They are then independent researchers, usually with their own reputation. The thesis is almost an afterthought."

Publication by doctoral students in the humanities, though less usual, is almost exclusively a sole enterprise, reflecting the individual nature of humanities research. The expectation is that once the doctoral thesis is completed a monograph may be published, or other publications might ensue. A history supervisor explained: "It's quite legitimate and common for students to go through . . . a PhD without having published anything. . . . It doesn't have a great deal of bearing on the thesis or its progress." In philosophy, it is regarded as "exceptional for [doctoral students] to do it. It's not just a matter of their capacity, but of the appropriate context of intellectual persuasions" (a philosophy supervisor). A professor of English literature commented: "They don't usually [publish during candidature], but if they do, we prefer them to do it alone. Those things are separate."

In theoretical areas of economics, as in mathematics and theoretical physics, students may submit collections of papers for theses, illustrating how theoretical knowledge fields across different disciplines share certain aspects of knowledge production. At the same time, they may have few characteristics in common with the more diverse experimental or applied specialisms in their own disciplinary areas. However, not all fields share such well defined norms, especially those which are newly established, transdisciplinary or subject to intellectual fashions and prone to the formation of cliques. Here the difficulty for students lies in identifying the relevant disciplinary foundations and adopting appropriate practices.

Sociology, for example, incorporates fields that have much in common with the humanities, alongside others that are more concrete and accretive, resembling science. Successful students have to navigate their way around the various expectations they may encounter in their departments; they have to go beyond departmental cliques to meet the expectations of an international community to which they may have only limited access. A supervisor in sociology made the point: "There are very different publishing cultures within the department. One colleague has just published a book as a result of twenty years' work, whereas there are other people who publish a book or an article a year. So there are a lot of different status hierarchies operating and it is very hard to make a statement that applies to students. But overall, there is a real shift towards what used to be regarded as an American 'publish or perish' sort of culture, and then again, it depends on what area you're in."

4. CITATION AND ACKNOWLEDGMENT

One aspect of learning the appropriate conventions for communicating with scholars concerns the unspoken rules for citation and acknowledgment, which are discussed more fully in a later chapter. In the humanities, the knowledge base requires that scholars assert a particularistic view that necessitates making strong judgements about existing literature. A supervisor in philosophy explained: "It is inevitable that a lot of [a doctoral thesis] will be quite explicit criticism about what other people have said. . . . It is judgemental in the sense of saying, 'you are wrong, I am right'." There is evidence that historians are less harsh than philosophers in their appraisal of others' work (see, for example, Becher, 1989b). Accordingly, the specific conventions of particular fields must be mastered if the doctoral student is to cite existing literature to the best advantage. There is a tension here which must be tacitly understood by doctoral students: they need to be confident enough to make the strong claims required to justify a new insight while treading carefully before potential examiners.

The tension between making confident assertions and treading carefully is not problematic for science students, especially those in laboratory or team settings, where there are many opportunities for learning the tacit rules of citation. The frequently collaborative nature of the publishing exercise means that others have a stake in appropriate citation and acknowledgment. Both the corporate nature of much research and the traditions of collegiality discourage strong judgements about the quality of existing research: the practice instead is that certain inclusions or exclusions are made in support of the work in hand. Students can glean how this is done in any of the research papers they read because the patterns are highly consistent and clear-cut. Should they err towards being too judgemental, their co-authors will soon correct the fault.

Learning the citation rules, however, is more difficult in specialisms where the knowledge base is less certain and the unspoken rules are not always clear. In these settings, argument is more individualistic and more dependent upon the assertion of theoretical and methodological frameworks, as exemplified by disciplines such as sociology or politics. Here students must be carefully attuned to the practices of experienced scholars in their particular specialisms, so that they may make critical judgements in appropriate ways. Bazerman (1981: 386) argues that the effect is to "reconstruct the literature to establish a framework for discussion." Unspoken rules for inclusion and omission in social science specialisms must also be deciphered, as Bazerman (1988: 283) points out: "Unlike articles in codified sciences where older texts have developed into the tacit assumptions of shared knowledge . . . the political science articles reassemble, reinterpret, and discuss anew wide ranges of the literature, dating back into the discipline's history." For doctoral students, learning these rules is a tacit process because, while they are highly field-specific, they are rarely if ever taught.

5. SCHOLARLY APPRAISAL

Scholars have a variety of available means for obtaining critical appraisal of work in progress, thus helping to ensure that no risks in credibility or status are taken when reporting research to the broader scholarly community. Established scholars use their networks not only to test and refine ideas, but also to reinforce the integrity of a fashionable specialism. Doctoral students have parallel and additional needs.

They need to understand the history of and the theoretical developments in their fields. In disciplines where research is individualistic, reiterative and interpretive, they must single out the contribution and context of significant scholars, together with the distinctive work of their own supervisors. In order to write an acceptable doctoral thesis, they must also express knowledge that embodies the disciplinary values of the scholars with whom they identify.

Students in laboratory and team-based science specialisms are at a marked advantage because of the proximity to members of their research communities and the high density and close communication networks of scholars in the broader scholarly network. Many are part of active and productive research centres whose funds permit them to attend conferences and maintain contact with key researchers. In some science fields it is standard practice to require students to present papers at seminars and conferences in order to connect them with the broader scholarly community and to provide invaluable feedback about doctoral work-in-progress. A geology professor whose funding base permitted this explained: "We put students in touch, people make recommendations . . . we encourage them to go to conferences here and abroad." Experienced scholars are accustomed to getting feedback on their work before it appears in the refereed literature. Students having to take part in these same practices usually find the process daunting but intellectually profitable.

A genetics supervisor described how his students followed established patterns of communication in the field: "If their findings are likely to excite interest . . . you might like to send it around and see if this excites anybody. About the easiest way to get a feel for things is to actually go to a meeting and present the work, by either a poster or as a spoken thing. And we encourage them to publish, it gives them . . . experience so they see the way things really happen." Research community meetings for sharing emergent ideas are a form of scholarly communication that opens up avenues for the appraisal of doctoral work in progress. A professor of agriculture explained: "We have a seminar series for each area within the school, which provides access to staff and students to look at other people's work. They learn from each other, but the students get a lot of good advice. It's something we can't teach. We make sure [doctoral students] give seminars. And I know everyone who twitches in the place." Students in this same department related how their supervisors seemed to have a close knowledge of where they were with their work, which gave them a considerable degree of comfort.

Students acknowledge the benefits of laboratory or team meetings in which their work is discussed or reviewed. In some cases, supervisors send student work-in-progress to significant scholars in the broader field for critical review. In fields where doctoral research is externally funded, there is a particular incentive for all members of the research team to ensure that students obtain scholarly appraisal of their ongoing work. In contrast, in theoretical areas of science, where research is more individualistic and the doctoral student's immediate research community is more limited, at least in numbers, students have fewer opportunities to learn from others in the immediate research community, giving rise in some cases to the need for supervisors to consult with established experts about their students' work-in-progress.

If the range of opportunities for obtaining scholarly appraisal is more limited in the theoretical, and therefore more individualistic, specialisms of science, it is especially so in the particularistic specialisms of the humanities, such as history, philosophy and literary studies. A history student described the process of seeking out appraisal during her doctoral thesis: "How did I know what to do? Well, we had to do seminars, which were pretty vague at first, but then you have the discussions . . . And then you have to find out who is working in the area. You might get their PhD thesis out, have a look at what's written . . . Then there were the jottings all through it [my thesis], and comments at the end. My supervisor would tend to write down a few pages of what she thought I was doing and how she thought I should improve it. It was good for me when I had friction [with others] because it made me think even more about what I was doing and how I was doing it."

Humanities students have to use their initiative to make opportunities for scholarly appraisal. As one reflected: "In Canada they certainly introduced you to appropriate people and opened up a wide arena for you. I've been introduced to professors at [other universities] through going to seminars and I am hoping to meet a lecturer from England . . . who is very much in my area." A history student asserted: "Here I felt quite lost . . . so I developed a colloquium with other post grads which I think really helped a lot of people . . .". Like their more experienced scholarly counterparts, humanities students can seldom talk about their own research in depth with other students or scholars because, owing to the nature of knowledge in these fields, nobody else shares quite their approach or topic. They need to be able to able to talk around their research areas with researchers other than their supervisors to get the additional affirmation on which they may depend.

There are, in the humanities, as elsewhere, notable exceptions. In archaeology, for example, there are core research areas concerned with generalities: collaborative research is frequent and the professional societies are strong. The research culture in which doctoral students find themselves resembles small-group science more than it does history, and students have as a consequence a wide range of opportunities available to them for obtaining scholarly appraisal of their work. A professor of archaeology commented on the typical activities in his discipline, all of which provide such opportunities. These include fieldwork, giving conference and seminar papers, talking over research with significant scholars, collaborating in the laboratory and joint publication: "All those sorts of contributions amount to something. And my students know what a research project looks like, they know what research questions are, they know what methodology is and they know their way around the effigies."

Since the research communities of the social sciences tend to rely heavily on conferences and seminars for canvassing information about research in progress and for disseminating the application of new techniques and methods, doctoral students in individualistic fields such as sociology are able to make good use of the available opportunities for appraisal of their work. Conference attendance tends to be supported, and it is usual for supervisors to encourage students to connect with other scholars in the immediate research community by presenting their ideas at colloquia. A politics supervisor explained: "It's a very isolated regimen that PhD students are left with now. Some . . . need to be told that what they are doing is worthwhile and pushed along, so we have found ways of getting them money for fares [to conferences]."

One of the difficulties for social science students is that, as previously noted, specialisms tend to be dominated by cliques. Students may find themselves in fashionable research areas or fields where internal divisions, methodological differences and strong theoretical divisions are common. While experienced scholars well understand the constitution and boundaries of cliques, doctoral students, without guidance, may not. This makes them particularly vulnerable in seeking scholarly appraisal, affirmation or direction from scholars in the broader research community. It is noteworthy that social science supervisors report more cases of conflicting advice given to doctoral students than is generally the case in other disciplines.

6. SOURCES OF CONVENTIONS

The published literature of individual fields comprises a key source where codes and conventions in specific fields are documented and made readily available, even if they are not explicitly stated. It is surprising, in view of the well established body of literature on disciplinary cultures, and the parallel literature on academic discourse, that these sources of disciplinary norms and conventions are not used more systematically as reference points for doctoral students.

A consequence of the lack of open appraisal and discussion of specific conventions for written work is that supervisors rarely, if ever, articulate for their students the principles underpinning writing in the field. It is a constant source of frustration for both students and supervisors that in the end, if the student fails intuitively to learn the appropriate principles, the supervisor will, to a lesser or greater degree, be pressed to show how they operate. Interestingly, this occasion for discomfort among supervisors is one that few feel qualified to confront openly. Taking the issue further, the discussion of discipline-specific discourse in doctoral theses in Chapter 8 puts forward a detailed analysis of the principles underpinning writing at doctoral level.

The primary responsibility for elucidating disciplinary conventions lies with the supervisor, or in some cases, joint supervisors, or supervisory committees. However, while this may be the principal source of advice about the development of the thesis, students must also learn from other contacts a range of disciplinary practices, values and attitudes, which must be reflected in the expression of their doctoral work for the thesis to be acceptable. The confidence apparent from the development of a disciplinary identity is critical to success. While some students expect their supervisors to be their key source of guidance, many assert the importance of establishing their own independence and the value of learning from a wide range of significant scholarly figures. This process can enhance the student's mastery of research methods and techniques that may be relevant to the production of a successful thesis. Not only can such considerations sometimes lie outside the expertise of supervisors, but they may also relate to debated issues at the boundaries of disciplinary acceptability, when students may well be advised to canvass wider scholarly opinion.

That nearly all supervisors were reluctant to describe themselves as mentors in the process of doctoral study suggests that the onus of assimilating disciplinary culture and conventions rests with the student. In research settings where collaboration and teamwork predominate, students are at a distinct advantage, in that the range of significant others in the immediate research community who can act as models or provide advice is relatively large. Furthermore, in applied and experimental settings, where external funding creates opportunities for doctoral students to attend national and international conferences, access to the broader disciplinary community is relatively easy; it is thus an informal requirement of progress to obtain critical appraisal in the broader scholarly arena.

In collective research settings, students are usually able to profit from the fact that the rules are shared. A genetics professor explained: "You've got a large group, where the student might not see the supervisor very often or for everything, but he or she has established the overall world view out of which the student is operating." A behavioural science student added: "In my area we have a seminar and I find that quite a good support group." There are particular lessons to be learned from co-authors, and more generally the research setting provides a rich environment for learning disciplinary values, attitudes, methods and techniques. A physics student remarked that his most significant source of appraisal was not his supervisor but another scholar in the department with whom the student had published: "He was by no means a senior academic in the field, but he was very, very knowledgeable. I found that he would come to me with an idea, or you would write a draft and he would give it his full attention, not like my supervisor. He would really think about it. He would give you comments almost immediately and it would really inspire you. And it was by no means that he was always right, I mean, there were cases where you might have conflicting ideas. But together you worked it out and you'd get a finished product that felt right."

The range of sources for learning disciplinary conventions and norms is wider for students in areas where team-based research predominates, and especially rich for those in settings where publishing, either with the supervisor or individually during candidature, is the rule rather than the exception. Though these are more competitive domains for students, there are benefits in weighing up advice from a range of sources in the wider scholarly field. A geology student described his tacit learning in the process of working up a chapter for publication after critical review from a conference: ". . . you try to think about content and you check around, re-examine and look for possibilities, ask some other people, get a feel for what the latest is. You might get an idea of what to look for and then you go back and pull it all together."

The situation in team-based science fields contrasts sharply with those in less cohesive areas, such as the social sciences, where the knowledge base is more fragmented and uncertain. There is not generally a high density of researchers in any given specialism, so students may have to depend heavily upon their supervisors for relevant external contacts. A politics student was well aware of the importance of his supervisor's role in seeking out important scholars to provide feedback on his work: "Picking a supervisor who has good contacts is important in terms of getting a good result [with the thesis]. They are the ones who can let you in on the latest developments, you know, the flavour of the month, or what's being researched. There are some very good students that don't get ahead because they have just not understood the framework that is currently being accepted."

An economics professor described the difficulties for students: "You have people [scholars in the field] who have got a wide variety of backgrounds, and they are moving into different contexts and applying those experiences. It doesn't work for students to try to get lost in there with all of them. . . . That's when I would step in and help a student to develop a background, maybe learn a new technique, read special books, write to somebody, meet a few people. You know, gradually to get some knowledge of an area." In social science fields such as politics, sociology, economics, commerce and legal studies, supervisors encourage doctoral students to broaden their contacts by undertaking part-time teaching, attending conferences on topics related to their theses, and participating in departmental seminars. Such activities are designed to enable doctoral students to connect with related researchers and to identify with a particular area from within which disciplinary values, attitudes and behaviours can be distilled.

In the particularistic knowledge fields of the humanities, supervision is most likely to be one-to-one. In many humanities fields supervision is described as a collegial arrangement, the understanding being that the supervisor, because of the individualistic nature of the research, serves as an experienced colleague offering guidance about the relevant conventions and norms. In these situations, where other scholars in the field are either limited or inaccessible to students, supervisors must assume much of the burden of providing a conduit to appropriate disciplinary behaviour. As a history student explained: "It's very hard to know what is wanted. . . . I suppose giving seminar papers and getting comments from other people have been worth doing, but I think the supervisor's input is really the only useful input I've had." The situation in theoretical science tends to be similar. A theoretical physics student felt that his department was a vibrant research enclave, but that everyone worked very much on their own, including himself: "You're looking to your supervisor's perspective on things. You see, I have no context other than that."

The allocation of teaching hours to doctoral students offers another opportunity for learning disciplinary codes of conduct and conventions. In the social sciences in particular, supervisors tend to extol the benefits of providing teaching hours as a way of inducting students into the research culture of the department: "Oh, we always try to get them teaching. They feel so much more a part of the department; it gives them status, and it helps financially." Teaching may provide incidental opportunities for socialising with experienced staff in the department, and it can also enhance students' chances of obtaining financial support for conferences and fieldwork, because, as part-time teaching staff, they may become eligible to apply for discretionary departmental funds. In social science fields where the importance of conference attendance is an essential means of sharing new knowledge, this funding support is vital. It is also important in humanities fields, where candidates need to compare methods, craft, perspectives and interpretations through dialogue with key scholars. Yet it seems that, in many universities, the available funds to support such networking are much more limited: Added to this, humanities students in particular lament the lack of teaching opportunities and the resulting incapacity to pay for their own conference attendance. The effect of this development is to limit the sources from which disciplinary lore can be learned.

7. KNOWING THE DISCIPLINARY BACKGROUND

The importance of acquiring a thorough understanding of the specialist area and of developing professional ties with significant scholars in the discipline cannot be overestimated. The first requires an adequate assimilation of the history and development of the knowledge base and its place in the broader discipline, which is derived largely through the study of existing research. The second requires recognition of the significance of key disciplinary players and an identification of those of like perspective. Association with disciplinary leaders can provide access to new, perhaps unpublished, knowledge developments in the field through informal communication channels such as seminars, conferences and general social contact. Both considerations are vital to career advancement and are the usual routes for established scholars; they also offer a similar range of opportunities for students.

But if considerations of eventual career advancement are important for some doctoral students, especially those in science and technology, establishing a presence within a field is a more immediate and crucial concern. A thorough understanding of the substantive area and its development is fundamental to the expression of knowledge at the doctoral level. Learning from those who are successful, and benefiting from the credibility that an association with them provides, are pivotal to acceptance by peers and to consequent professional success.

In areas where the knowledge base is densely structured and cumulative, as exemplified by the hard sciences, tacit learning acquired from important disciplinary players, as well as evidence of an association with them, are important levers to disciplinary acceptance because competition for recognition is fierce. In physics, for example: ". . . it is important, I think, to be recognised, especially early on. When you're early on in your career it's very useful to have the supervisor's name first on your papers – after all, he's the person [who is] eminent in the field, and people recognise where you come from." In geology, too, the mantle of the supervisor can help to build a student's reputation, as one student described: "Oh, coming out in the field and having a look at what everyone else is doing is really helpful. If they encourage you, that's very important and then you know [your work] is really good . . . it's got to do with the mind rather than the ability, and you start to feel you can be the same as them [successful scholars]." Doctoral students are keenly aware of the importance of associating with established scholars and of the benefits of being connected with their achievements in highly competitive fields where the students' acceptance is by no means assured unless they can demonstrate promise and patronage.

By contrast, in disciplines where knowledge is uncertain and less structured, as exemplified by anthropology or sociology, and where specialisms are liable to differentiation and internal dispute, it is important for doctoral students to establish professional ties which signal their intellectual affiliations. Where the concern is to enhance understanding rather than to add to the existing knowledge base, it is imperative to become identified with scholars whose methodological and theoretical perspectives accord with the student's own. In economics, a student who had dissociated himself from his supervisor was unperturbed: "I don't think that's any problem because I've had contact with those people who would be most important, both internally and externally, and I'm beginning to feel quite confident."

In the humanities, where knowledge is highly personalised and value-laden and where intentionality is important, identification with leading scholars of like mind can be significant on an individual level. Doctoral students have to establish a confident, individual identity through a sense of connectedness with their discipline, though often through a very limited range of scholars. Making contact them can be vitally important, as a history student illustrated: ". . . on an academic level, I think it certainly gives you a better understanding of yourself, because PhDs are so self absorbing and so internalised that you really start to question your own belief . . . somehow it can give you the desire for achievement, really."

While Becher (1989a: 55) argues that, overall, "the scholarly population has become too large, too diffuse to allow for the survival of a once-powerful patronage system", the findings of this study call that assertion into question. Formal patronage of students in areas of science, such as physics and microbiology, where the pace of research is fast and competition for intellectual territory is fierce, continues. Not only do supervisors assert the importance of "progressing good students" to advance their own reputations, but also students are well aware that some of their peers are given more assistance if supervisors think they show potential. A geology student explained: "It is a bit different for different students, like if they're really good they get progressed a bit," and, from a geology student "Yes, especially for them [the bright students], there are perks. I mean, one is that they go on all these trips. I'm going on one next week actually. Nice if you can get it."

For students in knowledge areas where competition for intellectual territory is not so marked, such as in the social sciences, and especially in the humanities, patronage exists, though it is less obvious. In the social sciences, supervisors attempt to ensure that good students are in line to be given more resources, such as the allocation of funds for conference attendance, especially if their work is strongly supported by a significant figure in the field. In these knowledge areas, informal patronage by disciplinary 'heroes' plays a vital role in the development of a scholarly identity and is a highly prized source of advice about academic rules and behaviours. The influence of disciplinary heroes is well recognised. Dill (1982: 315), for example, argues that they act as "powerful symbol[s] in time of stress and hardship as to the values of the enterprise essential to survival", and goes on to explain the importance of identification with disciplinary symbols who are real people. 'Disciplinary heroes', 'stellar performers' or 'gurus', as supervisors call them, are widely considered to be the beacons for doctoral students to learn about the standards and the major influences in the field. This view is also borne out by the views of students; for example a social science student remarked: "I met her when I went to England for a conference. She just gave me a kick-start. To think that somebody like her would think my thesis was going to be good, just made me keep going. And I was hungry for any hints I could get because she is so distinguished."

Science students, who by and large are able to identify more readily with scholars in their immediate research communities, seem to derive different benefits from their disciplinary heroes. In their status-conscious hierarchies, disciplinary heroes provide models of achievement and sources of academic conventions and conduct that cannot be overlooked by the aspiring student. An environmental science student explained: "Well, my supervisor's got various trophies for his research in his field, so he's quite famous. I'm not actually part of any of his trophies, but I hope I'm going to be, obviously, when I get a paper published."

8. THE SOCIAL BASIS FOR LEARNING

A range of resources is available in the typical research setting of doctoral students, from which the identifying features and values of disciplines and specific fields may be gleaned. Among the most important are those which provide the opportunity for students to experience at first hand how research is conducted and reported: both those that are explicitly set in place to ensure that the student learns disciplinary lore, and those that arise incidentally as part of the social setting for the research.

In the case of the latter, disciplinary values and norms are learned as a result of interacting with others in the internal, departmental research culture and also, more broadly, with the relevant scholarly networks in the discipline. In many cases, the values and norms are neither made explicit, nor do they form an overt part of the learning environment. Instead, they are represented in the contextual dispositions of the relevant scholarly community and are learned by tacit means. Bourdieu (1977, 1989) describes the contextual aspects of culture as the 'habitus', in which a system of durable dispositions are assimilated by members of a group as the basis for practice.

As students develop a more comprehensive understanding of disciplinary values and expectations, they build a platform of 'cultural capital' (Bourdieu, 1977), based on the accumulated wisdom of the group. In the context of doctoral study, this cultural capital is especially important, for it is through such inherited wisdom that disciplinary knowledge bases are maintained, developed and ultimately perpetuated. Yet while much cultural capital is learned explicitly, the degree of tacit learning involved needs closer examination. It is accordingly to the inexplicit processes of learning disciplinary values and norms that our attention may now be fully directed.

CHAPTER 6

COPING IN THE ARENA

1. NORMS AND EXPECTATIONS

Doctoral students usually have a good idea from their undergraduate years of what the disciplinary culture in their field is like. Those who continue in their first institution are expected to have developed some familiarity with the specialised writing conventions, codes of conduct, attitudes and values of staff and other students in their departments. Those who are new to an institution know there to be local as well as disciplinary rules to find out: a process in which they must use their initiative. Whatever their individual and disciplinary backgrounds, doctoral students are expected to undertake the typical work activities of their research cultures from the beginning of candidature. It is assumed that while students may need relatively more guidance early on and become more independent with experience, they will work under the same conditions as the rest of their research community. Beyond the general requirement to establish independence and the ability to 'drive the project', there is a range of expectations about achievements during doctoral study, differentiated by the values and conventions of their disciplinary setting.

Why is it that one student will be encouraged to publish as much as possible while another will be urged to devote all efforts to the completion of the thesis? And why do some feel intellectually alienated and emotionally alone while others yearn to be free of the shackles of their peers? The answer lies in the varied nature of disciplinary settings and is largely the product of the knowledge bases that sustain them.

This chapter examines those disciplinary features that are less explicit or which are altogether inexplicit, focusing on the means by which students learn them. The remarkable aspect of disciplinary differentiation is that some features are readily observable, while others are not. Nevertheless, students learn to identify key disciplinary features and adopt dispositions to act in accordance with them, even though they may not be fully aware of how these features play a part in conditioning social aspects of the production and reproduction of knowledge. The body of literature in the sociology on knowledge, particularly the work of Knorr (1981a; 1981b) and Knorr-Cetina (1983), addresses these social aspects of learning. Knorr Cetina's (1999) detailed analysis of the empirical, technological and social aspects of high energy physics and molecular biology exposed the lack of unity in science as well as the complex nature of learning to make knowledge in increasingly specialized fields of knowledge.

However, it is in the field of cognitive psychology that we can explore how doctoral students learn these empirical, technological and social conditions for making and reporting new knowledge. The interaction of explicit and inexplicit processes of cognition has been empirically investigated in this field. Trope and Gaunt (1999: 177), for example, found that "when situational information was made perceptually salient, accessible and applicable, it strongly influenced dispositional attribution" through unconscious or inexplicit learning processes.

Epstein and Pacini (1999: 479), accepting the weight of evidence from empirical studies, argue that unconscious cognition processes work interrelatedly with reflective, conscious ones. In Chapters 4 and 5 we have seen how conscious strategies put in place by supervisors and academic departments promote discipline-specific learning outcomes at the doctoral level. From the students' point of view, some of these learning outcomes are the product of unconscious learning as they build their disciplinary experience in the scholarly community. But it is also clear that some disciplinary conventions, values and traditions are tacit and must be learned by tacit means. To understand how these are assimilated, it is necessary to journey through the recognisable stages of doctoral study, to explore what the tacit values and conventions are and how they are learned. Reuben and Misovich (1999: 587) demonstrated empirically that "social constraints shape cognitive organisation", and that this becomes a reciprocal process in determining learning outcomes. In the context of doctoral study, such a process amounts to developing disciplinary savvy.

1.1. Being Savvy

Four distinct but overlapping kinds of disciplinary savvy that the doctoral student develops during candidature were outlined in Chapter 3. By way of review, these operate as reciprocal processes, so that social constraints shape cognitive organisation and cognitive organisation plays a role in shaping the social setting. The first of these key elements is grounding in the knowledge base – even though this may be multifaceted and drawn from diverse sources – and in its epistemological conditions – *grounded savvy*. The second concerns the ability to learn from and eventually promote and extend the values, aspirations and spheres of influence of the specialism – *cultural savvy*. The third is the capacity for learning from and contributing to the scholarly society through social interaction and the formation of scholarly networks. Through this process, termed *social savvy,* intellectual advancement and development are achieved for the individual and for the specialism. The fourth concerns not the making of knowledge, but its reporting. *Discourse savvy* involves mastery of the nature and structures of argument, etiquette for citation and acknowledgement, technical terminology and the nature and utilisation of specialised forms of metaphor.

A feature of these forms of disciplinary savvy is their interdependence. For example, it is difficult to imagine how other forms of disciplinary savvy might develop effectively without social savvy in the student's repertoire. Discourse savvy, accumulated throughout candidature by learning from the various formal and informal kinds of publishing and peer review, as well as from exemplars, supervisors and other students, is not easily separated out from the other forms. We shall now examine the interplay of the conscious and the unconscious learning processes, together with the interplay as cultural, grounded, discourse and social 'savvy' develops.

1.2. Allocating Time and Keeping Pace

One of the most fundamental strategies for doctoral students in the research process is the scheduling and allocation of research time in keeping with the typical work activities of the field, part of which is the commitment to meeting deadlines with appropriate outputs. Through these strategies, students come to terms with the typical methods and traditions of the field as they are plunged from the outset into the research culture and are called upon to exhibit a certain amount of savvy in terms of grounding in the knowledge base, its methods and traditions.

In team-based science fields, where research is frequently a collaborative enterprise that is externally funded and where the pace of research is fast and highly competitive, students are required to allocate their research time according to the constraints of the funding body and the targets of the research team. In physics, for example, the view typically is: ". . . if they're doing experimental work then they've got to know what's happening in our department and what our commitments are. We explain to them that after nine or twelve months they should really be taking the initiative . . . . We all come to work regularly. We see them every day. They know the drill" (a professor of physics).

Students in team-based research settings, as exemplified by experimental physics, geology and biochemistry, are accordingly expected to keep regular office hours and to have regular research group meetings to monitor their progress. A typical view from behavioural science illustrates this: ". . if I'm not in by nine on a work day, I give her [the supervisor] a call or something." The allocation of research time in experimental fields is not negotiable, from either the supervisor's or the student's point of view, and the expectation of steady progress towards fast-paced knowledge outputs impose a work structure on doctoral students.

In areas where research is a more interpretive enterprise there is less competition to produce knowledge outputs as quickly as possible. In the social sciences, for example, doctoral students are neither expected to keep regular hours of attendance, nor need they justify their use of time to supervisors. They are, however, expected to demonstrate that progress with grounding in the field is being made, particularly with the development of methods and skills appropriate to their research endeavours.

In all disciplines there is a general understanding, learned by students very early in their candidature, that there are different stages of doctoral research, each requiring different commitments of time and social activity. These stages manifest themselves differently in that they reflect disciplinary values and codes of conduct. Three overlapping stages are widely recognised.

The first stage is one of cue-seeking, where the student must be alert and sensitive to the disciplinary conventions, values and typical work practices in the department as he or she is socialised into a particular research culture. At this time, the student becomes aware of an overarching, international scholarly community, and begins to understand how specific conventions and work practices underpin the conduct of research in the field, and the way it is reported in the scholarly literature. Key values are manifest at this time; in interpretive fields, independence of mind is encouraged and rewarded, while students in science learn techniques and commit themselves to deadlines in a setting where there is little or no room for diversion or error. In this stage, cultural savvy is being established.

The second stage is one of creative development, where data is collected and analysed and its coherence and meaning are established. In interpretive fields such as the humanities and in many social science settings, this is a time when supervisors are in general reluctant to interfere too much or to be too directive. In science-based disciplines, this is a time when consistent outputs are monitored and results and interpretation are checked. Having already provided careful socialisation into scientific values and processes, supervisors now prefer to encourage intellectual autonomy and independence. Grounded savvy is developing and being shaped.

The third stage is one of slogging it out: by now the stages of enquiry have largely been completed, and a solid commitment to writing up the thesis must be made. This is the time of drafting and redrafting, of continuous refinement of writing, and the occasion for developing coherence among the different sections of the thesis. Discourse savvy, which ideally has been forming from the beginning of candidature, is now predominant.

Looking more closely at these stages in the interpretive fields of the social sciences, an economics supervisor elaborates: "Initially there may be lots of intensive consultations and then there would be periods where the student is pretty comfortable and can work on something for a while." A sociology student reported: "Well, I can just work at my own pace. There have been identifiable stages in my research, and for example, when I was collecting my data, I worked really hard. But in the beginning, I was sorting out a lot of issues, and I was more relaxed about that. And I think that was all right with my supervisor." In social science settings, progress frequently varies during the stages of doctoral work. The general expectation by supervisors that this will be so provides a flexible framework for progress that fits with the typical patterns of knowledge production. One danger to avoid, though, is that the flexible framework may become overstretched to the point of disadvantage by failing to notice problems of refocussing or refining the research, or difficulties of technical data collection or analysis.

One important consideration for doctoral students in social science settings concerns their progressive grounding in the knowledge base, and their perceptions about the nature of data. Many doctoral research topics involve large databases or utilise textual data that is cumbersome, complex and time consuming to transcribe, organise and analyse. It is therefore reasonable in the middle stages for contact with supervisors to wane somewhat whilst the demands of the databases are wrestled with. However, these demands occur at a point that is also generally considered to be the main period of intellectual challenge for students. As a result, many supervisors are concerned to keep a watch on focus and progress during this stage. A supervisor in management commented: "The middle stage is the major creative activity period where you need to give [students] time to let some ideas mature a little before you see them . . . . One of the problems is how fast it's going and one of the ways of ensuring you keep some control on those students is by maintaining contact, even if it be at the trivial level. They know and you know if they are not making progress at that stage, because that's when we often find ourselves quite hard worked; they want you to check on the workings they've done, or get them help with the programming, or they've got all sorts of questions about how they're going and so on. That stage should be quite intensive, because it's hard intellectually." These comments well illustrate the need for supervisors to be flexible in monitoring adequate progress, while at the same time setting parameters for making progress and ensuring that disciplinary savvy is developing. In addition, they highlight the importance of the supervisor having established early in candidature a close enough rapport to be comfortable about contacting the student if it seems that too little progress is being made.

The allocation and scheduling of time is different again in humanities fields. Here, ideas are relatively slow forming, owing in part to the 'top-down' nature of argument, in which the overall architecture of the research needs to be envisioned before the research can be written up. One outcome is that there are numerous potential impediments to maintaining a steady pace with doctoral research and reporting on it. It is widely accepted that students find the nature of the work intellectually and often physically isolating. In addition, many also find that it takes quite a long time to resolve the actual research topic. This is because, in order to do so, they have to adopt a personal theoretical perspective, and formulate an overall vision and architecture, which usually takes a considerable period of time. Here, the development of grounded savvy and cultural savvy are interrelated; the substantive knowledge base is not distinct from the values and theoretical influences in the specialism, and achieving sufficient mastery to progress to the next stage may be a protracted process. Taking some time to develop a focus can seriously jeopardise a student's intrinsic motivation to keep going. A history student described the well-known feeling among humanities students, of feeling lost intellectually: "I really had no idea what to start with. Eventually I found something that became very central to my thesis, and [then] I got organised because I knew what I had to do."

In the beginning, and sometimes into the middle stages of candidature, humanities students report that allocating time is not the main issue in making steady progress and feeling motivated to keep going. Instead, the prime concern is to achieve the necessary degree of focus and to develop confidence that the research is headed in the right direction. Here, the stages of starting out and creative development have to take place in tandem because the architecture of the thesis is conceptually constructed, as noted earlier, from the top down. Time allocation for humanities students seems to assume more importance during the final stage, when the writing up is being completed. During this stage, the pace of the research needs to be steady, with substantial amounts of time devoted to fleshing out the detail within the overall structure of the argument. Because effective thesis writing is the product of a process of refinement, students frequently take much longer to 'get it right' than they expect. Despite their own and their supervisors' good intentions, it is not uncommon for humanities students to overrun their allotted time span or period of candidature.

Developing disciplinary savvy and making good progress may be impeded by practical problems, too. One of the most widespread difficulties in keeping up the momentum of doctoral research results from students running out of money and having to work to support themselves while they finish writing up. A history student recalled: ". . the final stage was when I was writing up, and rewriting and rewriting and rewriting, and that was twelve hours a day, seven days a week. I also started teaching during that period, when my scholarship ran out, but tutoring involves heaps of time . . . (while) you're trying to do the PhD. I don't know how people do it for the whole candidature." Another commented: "It is so hard when you have your head in one place, but then you have to switch it to another place because you have to earn an income. It can take ages to shift emotionally and intellectually." These concerns are amplified for science students, especially those in applied or experimental fields where the pace of research necessitates the expeditious completion of theses because findings can quickly become dated: "That big drop just when you have no real assured income . . . I didn't really know where the next dollar was coming from" (physics student), and "You just seem to be scraping and scrounging and worrying about candidature all the time. You waste time while you're earning money somewhere else, basically" (geology student).

1.3. Recognisable Stages of Candidature

In the first, cue-seeking stage, students in interpretive fields have to develop cultural savvy, to come to understand scholarly values in their fields. They also need to develop their grounded savvy by familiarising themselves with the knowledge base, its traditions, methods and approaches -they have to map out the intellectual landscape, the key players, disciplinary standards and a wealth of conventions governing the ways in which new knowledge is reported and disseminated. During this first stage, students report a need for solid commitment from their supervisors, including strong intellectual support and encouragement to keep up the momentum of the research, especially if there are unresolved issues of focus as time draws on. They also comment on the role of the supervisor as a key source for learning appropriate values and conventions.

A very different scenario typically occurs in science, where social savvy is essential, in that science supervisors are primarily concerned with the 'neatness of fit' of prospective and beginning students. Students begin doctoral work equipped with a research topic, a comparatively narrow focus, an idea of deadlines and commitments in the research program, and close direction towards scheduled progress. Cue-seeking here requires, at least initially, relatively little initiative from the student, who has to become rapidly socialised into the ways of the department, including its values and aspirations. Typical work patterns and techniques, writing conventions, scholarly values and imperatives, and how to fit in with others in the department, are readily observable in the research environment in which the student is an active but junior member.

The second stage of creative development is something of a long haul for most students, regardless of discipline. However, the typical activities of this stage vary across disciplinary settings. With topic issues resolved, it is the time for producing results from experiments or for amassing materials and developing the argument, processes which may involve considerable writing. For social science students it is frequently a time of demonstrating deft analytical ability and checking interpretations of data, which can take some time to complete satisfactorily. A common thread during the creative development stage is that students assert their intellectual independence. Disciplinary savvy is by now expected to be manifest. For humanities students, this can be a period of vulnerability and low confidence. For science students, tentative intellectual steps are made and checked by the reactions of colleagues and supervisors. Having learned and adopted appropriate values and conventions, the student is now expected to use them, and in doing so, to contribute to the evolution and development of the knowledge base.

The final stage of slogging it out is concerned with completing a coherent draft and expressing disciplinary savvy in the thesis through mastery of the specialism's discourse. The process of building up a coherent and sustained argument usually requires many cycles of review and rewriting, refinement processes which produce quite different discourse structures according to the nature of the knowledge base with which the student is working. In the next chapter we discuss more fully the relationship between disciplinarity and discourse, but for the present it is important to note that the student in the final drafting stage is attempting to master – and further, to manipulate - the writing conventions that are accepted in the field. Progress here is likely to require a more steady pace than earlier stages, and this can be difficult for students to sustain at a time when the initial enthusiasm for the project may have waned, ad a student in education remarked: "I'm just going through the motions. All I want to do is to be finished with the thing. I want this to be over."

The variations in work activities of each of the three stages give rise to marked differences in the typical roles of supervisors in particular, but also of scholarly mentors and significant others, in providing sources of extrinsic motivation and affirming forms of disciplinary savvy. In humanities disciplines, the relatively isolating nature of research means that there are fewer people available to be mentors, significant others or critical friends.

As a result, supervisors play a large role in encouraging students to sustain the momentum of the research in circumstances where deadlines are usually artificial and keeping up the pace and maintaining intrinsic motivation to continue can be difficult. During the final stage, students in collective research areas clearly have an advantage: they can continuously check and compare their concerns, attitudes, approaches and achievements with others in the research area. Students in the humanities typically describe an important intellectual affinity with their supervisors that is needed not only for intellectual support but also for encouragement and commitment to keep going with the thesis while life's distractions continue unabated. One student reflected: "I've had major emotional traumas because my partner and my friends, none of them are academic at all. And you really have to isolate yourself. You can't even mix socially with people who don't understand your problem that you're working on. I'm an academic here and a fiancÈe there and a best friend on the other hand, and they really don't gel . . . then the other side of me says, well, I love history and . . . with [my supervisor] I sit down and have a cup of coffee and just chat with him on a one-to-one level. There's an openness and an ability to relate."

1.4. Supervisory Relationships

The nature of the supervisory relationship is widely acknowledged as being especially important to success; Youngman (1994: 101), for example, noted that experienced supervisors maintain a strong guidance profile while engendering high levels of independence. While there is a wide range of individual relationships between supervisors and students, the present investigation shows that the constant element is the supervisor acting as a touchstone with disciplinary lore and as an adviser through whom the student can make important contacts and learn the ways in which disciplinary recognition and credibility are earned. This is especially important when the supervisor is the main source of contact with the discipline, as is often the case in the scattered specialisms of the humanities. As a history student explained: "Well, one of my supervisors is very formal, very academic, and the other, well we have lunch and she's more like a friend. But in the end, it's how they are connected and the sources they guide you to and how they keep you informed. Those are the things that supervisors contribute that keep you going."

Supervisors employ different techniques for urging their students to keep up a steady work pace in the absence of external pressures to meet deadlines. Another history student described how her supervisor had encouraged her both personally and professionally: "I think the pressure is a bit heavier now [in the third stage of candidature]. In the second year you get the two-year doldrums and a lot of PhD students go through the blues, wondering if they're doing the right thing. And I went through it . . . . My supervisor was great. He'd say, we've got to do this and this and this. He said 'there's this postgraduate conference coming up next year. Maybe you should do a paper'. So he just phoned up and booked me in. At the time I thought, 'My God, what did he do that for?', but it was actually the best thing because he pushed me and made me do it . . . in my case, the supervisor sets the deadlines." Other supervisors use different techniques, such as imposing artificial targets, checking work diaries or setting regular meeting times for which written work is always required. Pressure to show disciplinary savvy in the arena, such as presenting at seminars to established scholars, tends however to be the most effective and relevant strategy.

Many humanities supervisors argue that, to compensate for the uneven pace of doctoral progress, it is necessary to forge reasonably friendly terms with students from the beginning. They reason that in this way, their students do not feel intimidated if they are contacted about slow doctoral progress or if a deadline is suddenly imposed to get the student working productively again. Indeed, some supervisors in the humanities feel that reasonably friendly relationships are essential to ensuring productivity in doctoral work and that engendering personal commitment is a valuable technique.

In contrast, the relationships between supervisors and students tend to be more formal in science settings, even in the context of theoretical physics. In foundation areas of science, where scholarly hierarchies are traditionally rigid, supervisors adopt an almost casual attitude towards their students: "We have meetings and we know where they are up to . . . . Without being brutal about it, you might let them flounder about for six or nine months, but that's it. And they usually sort it out." (a physics supervisor).

An important difference between theoretical physics and humanities students is that, as students in the former field are generally expected to be at work in the department every day and to meet set deadlines for the production of results, supervisors can easily gauge the progress being made. Many science supervisors do not consider that a friendly relationship with students is necessary at all, and certainly not as an aid to checking informally on progress. Unfortunately, students sometimes interpret this attitude as indifference or as a lack of commitment to their doctoral project. One student spoke for his peers in stating: "Outputs are so important. If you don't have individual meetings to spur you on, then you can end up with the idea that there seems to be a lack of interest in your progress." Here, the student echoes an imperative articulated by Becher, Henkel and Kogan (1994: 72): "the dominant purpose of those in charge of knowledge production . . . [is] . . . to get the work done at the speed required to keep pace with change and the competitive field." At the same time, the student must be aware that the way to get on is to produce outputs, regardless of personal considerations.

In areas where knowledge is concrete and accretive, and research questions can be broken down and tackled in smaller chunks, the pace of research is fast and characteristically competitive. There is, therefore, pressure to publish results as soon as they are available. This brings other advantages. Publishing during candidature assists doctoral students to mark out intellectual territory, and provides them with a series of deadlines for their research that enforce a steady pace in yielding results and writing them up. Not only are students forced to meet personal schedules, but they are also subject to deadlines that may be externally imposed by funding bodies, as well as those imposed within their research teams. The individual contribution of a doctoral student's work to the greater team effort is often checked through the progress of the research team. A genetics supervisor explained how this works in practice: "How closely the supervisors and the students work varies. There are people who have several hours' interaction with their students every week, some have day-to-day interaction, and you'd see everyone during the week if you could. But typically the supervisor is working in the same laboratory, so there is that presence, and though they might not be doing bench-type work any more, they know where everything is up to, they have to." Reflecting the hierarchical nature of science research enclaves, another genetics supervisor commented: "Close personal relationships don't occur in this department, or in this field, to my knowledge. Here, people keep a professional distance from their students."

There are many notable exceptions to the notion of 'professional distance', of course. In biochemistry, for example, the practical nature of team-based laboratory work and the rewards of external funding for successful outputs ensure that researchers make strong professional commitments to each other, and that supervisors do so to their students. One distinguished professor described the setting in his department: "It's close, you get very close to nearly all your students. Some of them don't allow you to get close, but I mean, you share the most important things in life, all the aspirations for their future . . . you share their disappointments. We have terrible disappointments in this field. Eighty per cent of the experiments don't work . . . and so you see the students go through absolute troughs and valleys. And when they do have a high and they're jumping out of their skin, the first thing they want to do is tell you. Share it with you. You'll often go and get a cake or a bottle of champagne. But you find this in most labs in the department. I mean, you see every lab leader having their little parties. They socialise with each other, but oh, it's pretty professional in our department, I think."

In the social sciences, where the boundaries of knowledge are not well defined, and where there is relatively less pressure to meet deadlines, new research questions can be taken up in a more *laissez faire* manner. In fields such as sociology, economics and politics, doctoral students may take some time to negotiate the research question, and progress during the different stages of candidature is not likely to be consistent or steady. Keeping pace is seen largely as a function of being clearly focused (Parry, Atkinson and Delamont 1994), even though supervisors expect the progress of doctoral work to vary according to its stage: "In our field [economics], a project can be tailored to three years' full-time work. Typically the student is thinking of taking on more than can really be got through, and as time goes on, it gets more laid down and things become more focussed."

A supervisor in legal studies crystallised the difficulties of keeping up a steady output during doctoral research in an area where the knowledge base is uncertain and where theory provides a general framework for understanding. A law supervisor explained: "sometimes our [doctoral] students might seem a bit slow on the uptake - it just varies. It is hard to do the degree in three years anyway, but if they are doing something really unusual or inventive, like that one recently about the phosphate mining and exploitation, or the one on the peace army in India, it is hard to monitor the pace because there are so many blockages, both intellectual and practical. So you can't generalise." Supervisors in these settings nevertheless understand that students need encouragement to remain motivated and productive, and they frequently describe having a key role in helping students to narrow their focus and remain confident, in the face of interruptions to progress and uncertainties in the knowledge base. While supervisors may not openly discuss blockages in the flow of activity with their students, their role is shaped by the need of doctoral students for affirmation of work-in-progress, regardless of pace or consistency. Supervisors in these fields frequently comment on the importance of encouraging students to present at colloquia and seminars to ensure continuity and commitment, and of finding doctoral students part-time teaching work in the departments as a means of assisting them to develop disciplinary savvy.

2. MASTERING THE DISCIPLINE AND THE FIELD

In Chapter 2, the heterogenous nature of the work activities, values and the identifying characteristics of disciplinary fields were described, and the impact of dynamism and change in specialisms was discussed. This dynamic, evolutionary feature of disciplinarity is an essential part of the knowledge that doctoral students must acquire. They must learn to identify how their research is situated in relation to the work of others in the field, and they must be able confidently to assert that their own contribution to knowledge is worthy. That this is an objective seems rarely if at all to be made explicit to students, yet supervisors convey this message to them in a number of ways. A history student explained: "I guess I fall more into a labour history camp than I do into lots of other strands. And that camp I see as pretty corrupt and some of the work quite ho-hum, [but] there's some interesting work. I guess you find with most camps like that, there is a lot of attention devoted to questions that I think are . . . a bit pathetic and not worth buying into. But those issues seem to split people into, 'these are the ones who are in', and 'these are the ones who aren't'. Do you know what I mean?" Typical supervisory behaviours here would be to encourage the discernment of different schools of thought and scholarly cliques and the student's development of a position and value frame in relation to them. Usually supervisors describe this process simply as "narrowing the focus" and yet tacit norms and counter-norms are clearly being navigated in the relationship between the knowledge base and the community that sustains it.

A supervisor in law gave more insight commented: "Well, human rights law takes you into so many other fields; there's public policy, politics, sociology, cultural studies, various related fields of law like migration. It's very hard for the student. Usually they've already got an interest in a particular field and that becomes the focus. Even though there can be tangents off that. So I guess the other limiting factor is to keep in mind who the examiners are likely to be." Not only is it important for the supervisor and the student to check the viability of their focus by presenting at conferences and the like, the list of potential examiners also provides parameters.

Narrowing the focus of the thesis appears to be a different proposition in science, where the research question is clear and it remains to be resolved how much evidence is sufficient to make new knowledge claims. In contrast, narrowing the focus may be especially complex in fields where the knowledge base is not concrete, as in the social sciences and humanities, or where the area of study is transdisciplinary.

2.1. Dealing with Uncertainty

In transdisciplinary fields, students face difficulties of identifying the relative contribution of distinctive theoretical frameworks from previously unrelated fields. They also have to learn to make appropriate and acceptable links across the fields for themselves. A student in education reported: "There's quite a few angles to my work, and I got diverted on all sorts of tangents for a couple of years. There wasn't anybody to point them out to me. I've got an eclectic background, degrees in this and that. That probably made it harder . . . eventually I met someone who said, 'look, you've got to adopt one angle and chop off a lot of the not-so-related stuff'. It's been a lot easier since then, but what do you do when there isn't anyone, really, in your field to supervise you?" This student's comment echoes a related phenomenon. Parry, Atkinson and Delamont's (1994) study found that identity and work issues are more problematic for multidisciplinary students. They showed that doctoral students in these circumstances could not easily relate to one body of knowledge, which in turn affects the focus of the thesis. Here the comments above of the law supervisor are telling. With so much uncertainty in the knowledge base, and without clear disciplinary parameters to guide the development of a theoretical framework, other parameters need to be found. These are the examiners, whose interests and values come to play an important part in the success of the thesis.

In social science fields it may be difficult to provide adequate supervision to cover the different aspects of doctoral research topics, given the wide range of possible perspectives, topics and methodologies. It should be a cause for concern that for some years researchers have documented the relatively limited research supervision experience among social science supervisors (see, for example, Youngman, 1994; Buckley and Hooley, 1988). The trend towards increasing enrolments in doctoral programs in many countries, together with the limited supply of experienced supervisors, suggests that students in social science fields are faced with many uncertainties. They especially need to use their own initiative in both the immediate research environment and among the wider scholarly network if they are to learn values and conventions appropriate to their fields, and at the same time remain alert to the changing constellations of influence in them.

If social science students face these challenges, then the challenges faced by students in transdisciplinary fields, or the "new modes of knowledge production of 'post normal science'" as Weingart (2000:36) calls them. However, he goes on to add, ". . . the empirical fact is that the 'real problems' are constituted by existing knowledge and its gatekeepers" in effect arguing for better understanding of these two parameters of knowledge making.

2.2. Publishing and Peer Approval

The rewards of publishing offer professional recognition and the incentive to keep up a steady pace during the doctoral program, thus providing an important extrinsic motivator for students working in fields where competition for intellectual territory is keen. Students in these settings learn that obtaining validation of doctoral work in progress from the broader disciplinary arena is a necessary process, and that publishing is an important vehicle for doing so that can supplement the advice of supervisors. In mathematics and theoretical physics, for example, validation via publishing is a means by which both the doctoral student and the supervisor obtain feedback about the acceptability and likely impact of doctoral work from influential scholars in the field. Positive appraisal through publishing gives students a recognised role in the process of producing knowledge, and establishes their identity and intellectual territory in the discipline. In the highly competitive fields of science, publishing during candidature is never considered a trade-off against candidature time. In an apparently increasing number of scientific fields, collections of papers can be submitted as theses, a process which guarantees ultimate thesis success. A mathematics supervisor attested: "If the damn thing's essentially a bound-together set of related research papers, it isn't terribly difficult for them to write well . . . and anyway, most of our students have had three or four papers and (so) they've been off

and written the bloody thing." Hourcade and Anderson (1998) refer to this phenomenon as 'opening doors' through the publication of acceptable contributions to
knowledge.

Where team-based research predominates, as exemplified by chemistry, genetics,
and microbiology, publishing provides an important validation of and encouragement for work-in-progress. In these fields, publishing is instrumental in the process
of socialisation to the professions, as described long ago by Becker and Carper
(1956). Socialising in fields such as these involves developing an understanding of
the nature of characteristic problems, adopting a genuine interest in them and
becoming confident in the expression of knowledge in the presence of experienced
scholars. A professor of microbiology described the conditions for socialisation to
his field: "If you are doing good work and you're on track, you should be able to
publish, and we do that [urge students to publish] as they go through, so they stand
up in the wider community . . . we send them to conferences as well, and they meet
people in the area and present their work and get introduced to the wider community
of microbiology. They get MASM and FASM after their name, and that's very important." An experimental psychology student commented: "Well, it's . . . the ultimate
reward, I guess, apart from getting your PhD accepted. It's very important. I mean,
I've had some of mine published, you know, and it's just a great feeling. It just felt
like it made a lot worthwhile, I guess."

In contrast, in social science fields as exemplified by sociology and politics, there
is little competition for claiming intellectual territory, and research questions can be
pursued according to personal interest. As a result, there is generally less pressure to
publish during candidature. In many areas, effort put into publishing is seen as com-
promising candidature time, though there is always encouragement for very able
students to publish, since marking our intellectual territory cannot begin early
enough for those likely to be 'stellar performers', as one supervisor called highly
successful students.

There has generally been limited competition for intellectual territory in the
humanities due to its highly individualistic nature, although Henkel (2000) noted
that pressure to compete for high profile territory appears to be increasing. In these
fields, where knowledge is highly interpretive and individualistic, it is more usual for
the thesis to be completed and for that to form the basis of subsequent published
works. The increase earlier noted in transdisciplinary research in humanities and
social science fields may, however, precipitate greater pressure for doctoral students
to publish during candidature as a means of achieving scholarly acceptance for
research that is novel or which transcends accepted disciplinary parameters.

2.3. Benchmarking Progress

Publishing during candidature is not the only means of obtaining scholarly approval
and validation of work in progress, nor is it always appropriate or possible. However,
the need for critical appraisal, validation and affirmation, as well as for keeping up to
date with the knowledge base, are vital to success. When doctoral students are able to
network through department-based seminars and colloquia, and to participate in

meetings of the broader scholarly community through conferences and the like, they have opportunities to benchmark their approaches, theoretical frameworks, methodologies, techniques, intellectual dispositions, knowledge of recent developments, scholarly connections and research progress with other scholars and students in the field or in related fields.

Though it is a less than conscious, tacit process which is difficult to define and articulate, it is nonetheless vital that doctoral students compare and contrast their intellectual performance, approach and standards with others, all the more so when their research is interdisciplinary. Understandably, benchmarking is especially important in fields where the knowledge base is abstract rather than specific.

In social science fields, where knowledge is highly theory dependent, and where technical and theoretical developments are frequent, scholarly appraisal of doctoral work-in-progress is particularly important. Presenting papers at conferences, delivering work-in-progress seminars and participating in colloquia are all avenues for obtaining scholarly appraisal, and they provide opportunities for students to become familiar with the research and influence of established scholars and for finding out about new developments. An economics supervisor explained: ". . . here there is an emphasis on keeping up to date, because of quick changes in the field; that's very much the case in economics. You can be quite out of date if you don't read the latest journals and if you don't mix with people."

In fields where research communities comprise ideologically divided cliques, as exemplified by politics and sociology, doctoral students learn that it is the norm to identify with a school of thought and to observe the conventions and behaviours of a like-minded community of scholars: "Sociology is such a diverse discipline theoretically that it really would be very problematic if you actually didn't let students make judgements about who they relate to, and which theorists they will use" (a sociology professor).

3. DEVELOPING IDENTITIES

The need to identify with congenial networks of researchers who share similar theoretical frames of reference and methodological approaches is also manifest in the field of education, where specialisms may have their foundations in one or more of a wide range of disciplines. Since doctoral topics in these specialisms are often derived from practice, it is important for students to identify with and master disciplinary and theoretical perspectives that are appropriate to their research and that are likely to be acceptable. In these circumstances, feedback from like-minded scholars is especially valuable. An education student explained: "I knew what I wanted to research but I didn't know what the theoretical developments were in the field. Actually the field is very broad, so I had to find ways to narrow it down. I did that by honing in on the work of a few key people and by getting to know some of them, and eventually, getting feedback from some of them." Feedback of this kind is especially important because, in settings where there are diverse ideological differences amongst groups of researchers, students need validation not only of their interpretations of data but also of their perception of the relevant phenomena and of the appropriateness of their methodology.

While it is important for doctoral students to have opportunities for obtaining appraisal of work in progress from the broader scholarly community, a pressing problem in the social sciences, and also frequently in the humanities, is that such opportunities are limited. Funds are usually constrained, and the diversity of research interests sets natural limits upon the range of research areas catered for by visiting scholars and fellows. A supervisor in education in a relatively large university described the difficulties in his context: "Our staff represent the range of disciplines in the university, right across the range . . . there is not a lot of opportunity for publication. We have an allocation for them [doctoral students] to go to conferences, even though the chances are they won't get it. If an important scholar gives a paper here, there is usually a small [audience] showing because people here have different backgrounds. We don't have the critical mass. One thing we have set up is to have small groups, communities of students and . . . (sometimes) . . . supervisors who meet regularly, support each other and know each other's work, so they are offering politicality and peer support. I personally don't think we offer a very conducive environment for postgraduate study here. We just can't afford it."

The opportunities for doctoral students to benchmark are also limited in the humanities, where students report feeling a sense of impoverishment as a result of financial restrictions and of small, scattered scholarly networks. In these often impoverished settings, however, it is considered essential for doctoral students to develop their individual identities, for which networking with relevant scholars is an essential element. In fields as diverse as American history, logic, English literature and religious studies, supervisors report not wanting to 'interfere too much' with the work of their students. To offset this, supervisors seek to confirm the appropriateness and acceptability of doctoral work using relevant outside sources. Students seem resigned to this situation and the implications of it, as a history student explained: "I think it's up to you, you've got to get out there and find it out for yourself. But in these practical things which really make you a part of the department and part of introducing you to the academic world, just in very practical ways, this department really was very unhelpful . . . it's nobody's fault, there's absolutely nothing [for students], there's no money whatsoever." (It is significant that in a different context, the sense of impoverishment is especially keenly felt by students in theoretical physics, who constantly compare themselves with their better-funded counterparts in experimental fields).

Because humanities research is highly individualistic, researchers find it quite difficult to share core elements of their work with their colleagues. The level of intellectual exchange at conferences and colloquia may, therefore, not assume the same importance in terms of keeping up to date with new developments in theory or methodology as it does in other disciplines. In settings where the international networks are small, eclectic and scattered, it is easy to understand the multifaceted nature of the isolation expressed by humanities students, and the resulting difficulties for them in effective benchmarking and in keeping motivated and maintaining momentum for the long period of candidature.

3.1. Relating Intellectually

Against this background, humanities students clearly identify three sources of motivation to keep going with their doctoral work. The first is the commitment and encouragement of the supervisor. The second is positive appraisal of work in progress from exemplars in the specific field. The third is encouragement from fellow students with whom they can 'talk around' their individual approaches to research and their personal difficulties. A mature age history student elaborated: "You need to get away and talk to somebody else, you feel so isolated, head down. My supervisor's attitude was, 'oh, nip off to England!' They just don't understand about having no money, they think you're whooping it up on a scholarship! For me, there was a group of two or three other students. Without them, I probably wouldn't have finished it, because they've got the same sorts of problems as yourself and you need them. They were absolutely essential." This student went on to explain how she could benchmark her progress and methods with those students, and to a limited degree, her discursive abilities, but it was difficult to have coherent discussions when one discussant knows nothing of a topic area.

What roles do supervisors and the range of 'significant others' play in shaping the ability to undertake and report research at the level of the doctorate, in ways that are acceptable to and appropriate in particular fields? Across the range of disciplinary settings, from the laboratory team in experimental science to the student meetings in history, students describe their need to benchmark their progress, their approaches to research, their skills, techniques and values, with mentors, critical friends and a range of significant others. More importantly, they are learning by tacit means the conventions and the norms that shape acceptable research and appropriate reporting at the doctoral level. The key role of the supervisor is very often to identify a pool of 'significant others' so that benchmarking may take place.

3.2. Learning the Ropes

Nowhere is effective benchmarking more keenly reflected than in the writing of the thesis. Overarching epistemological issues, such as the nature of argument, questions of style and features of language, will be discussed in more detail in Chapter 7. First, though, the kinds of strategies that lead to successful completion and the kinds of advice supervisors and students find useful are worth examining because they illustrate well how the nature of the knowledge base prescribes the values and conventions for thesis writing in different fields.

A geology student whose doctorate involved team-based research described his strategy, directed by his supervisors: "The beginning part would be when you're doing a lot of reading, and you're getting out watching and working in the field area, going out and poking about and seeing what's out there, doing a bit of a tour of the countryside . . . . I think it's a continual phase really, because you do your beginning background work and then you get into it a bit and then you discover you didn't do enough . . . and you go back and do some more. You begin with the literature review, but you have to write up all the time, from after the beginning phase. You have to keep going back and checking what you've got with the literature, again because . . . you have to show you have the support of what's already been done . . . There's so much speciality involved with what we do that if someone does not take them under their wings when they're working in their middle stages, when there's that acquisition and the interpretation, there is no control or check out for quality . . . If anything, it's a triangle and everything funnels down to this one point. But it's very practical. In fieldwork you have to keep going back to your area to have another look. There's this process that when you've got enough evidence, you know you're close to the end and there is a light at the end of the tunnel. You can't believe it." This student clearly describes writing norms that relate to the accretive nature of science, and concludes by describing the structure of the argument that, as remarked earlier, is built from 'the bottom up'.

A different strategy is to be found in history, where the knowledge base is less certain and the writer's perspective shapes the argument, which is constructed holistically, 'from the top down'. A full-time student described how she wrote her thesis: "In the beginning, the seventeenth century was a really 'history' thing to do, whereas now it really isn't particularly concerned about doing the history thing. When I was writing, [my supervisor] would say, 'Oh, I see what you're doing, but what about these issues?' And you'd say, 'oh, yes, I need to address that to make sense of my argument, or whatever . . . [My other supervisor] put comments on one chapter, which sort of annoyed me. In one sense I found it very intrusive . . . I only properly started the writing up when I knew basically what I was doing. It took two years including the time overseas. I had a lot of material assembled and I knew what I wanted to use. Then it took me a year to get it all finished and that was twelve hours a day, every day of the week, and I saw the supervisor a lot on the written work . . . she was really good, she would really try to attend to what I was saying. She would read and re-read it and re-read it and she would put in comments."

A contrasting strategy may be seen in the social sciences, where a student described working on a thesis based on grounded theory: "My undergraduate degree is in science. I fell into this field. So I did things in order - the literature and then the data collection, and then the writing up. I even published things from the database. But I couldn't get out of writing up the literature chapter. It was a nightmare. I never finished my literature chapter! Then I got a new supervisor who was very experienced. He just told me to forget the literature. I was wasting my time. He told me to write up the data chapters first, and then develop the literature according to what I had. And that's worked well for me." This student's comments illustrate the importance of developing a working knowledge of the epistemological foundations of the field and aligning the research and writing-up strategies accordingly. It also illustrates the theory-dependent nature of social science research. The student could not make adequate progress until the theory driving the argument had been articulated: in this case, grounded theory. Of course, a wide range of research methodologies and theoretical perspectives are employed in the social sciences, so effective strategies for writing up will vary accordingly. It is the recognition of field-specific conventions that students must achieve.

4. INDUCTION TO KNOWLEDGE COMMUNITIES

In their reflections about the critical events that increased their motivation for completing their doctorates, students from all disciplines acknowledge the importance of receiving encouragement and approval from influential scholars. In some cases there is patronage from an exemplar in the field. Patronage, however, is highly differentiated across disciplinary cultures. A more detailed account may be found in Becher (1989a). In law, it seems that the way to get on is to have a mentor whose area of interest is followed up by the junior colleague. In history, the best scholars tend to attract the best students; the more extensive the range of contacts the supervisor has, the more influential he or she is likely to be. Students understandably attribute added value to patronage or sponsorship by significant scholars. A physics student, whose research culture was explicitly hierarchical, commented on the status of a professor: "His students really try very hard to impress him. They set out to impress him." A geology student whose supervisor had an esteemed record of attracting large external grants stated: "I sometimes think I am very lucky . . . I think it's fantastic being involved in something that's very, very enjoyable to you, and having the honour . . . of being adopted by your supervisor."

One of the most obvious roles of exemplars is that they model the norms and values of their research communities, so that students learn from them and are guided by them. A student in experimental psychology described her supervisor, an international expert in her field, as her 'guardian angel', because her status meant that her advice was authoritative while the connections with key researchers she was able to provide were invaluable. Another student, this time in Australian history, had forfeited a scholarship in her home city in Canada to go to Australia to work with a significant scholar in her field, even though it cost her $9,000 per year: ". . . the reason why I did that was to work with [the supervisor]. I'd used a lot of his work during my master's degree and I discovered he was here, so I wrote to come here and here I am. And he's definitely worth it."

When students have the opportunity to network with exemplars in their fields and learn from them at first hand, there are usually rich dividends. Those who can do so have a special advantage in being able to identify standards, values and disciplinary aspirations with ease. Exemplars are also disciplinary gatekeepers who can provide advice that has special authority and that engenders confidence in doctoral students. Though Whitley (1984) describes the significance of exemplars in setting standards and directing strategies as varying according to the nature of the discipline, there is no doubt that benchmarking with and gaining approval from an exemplar in any field provides a level of confidence not easily obtained from other sources. Just as established scholars may seek advice from exemplars about their developing ideas, so may doctoral students, who have so much more to gain from the process in terms of disciplinary acceptability. A politics student explained: "I know that if I acknowledge the advice I've got [from an international expert], then my thesis is likely to have a smoother run when it gets examined. After all, he's the absolute top of the field and nobody in their right mind would question the content or style of my thesis if he's already approved it. And in my field, style is a fairly rarefied thing."

4.1. Learning Diplomacy

Gerholm (1990) argues that an important form of tacit learning for graduate students concerns diplomatic conduct. This is especially important because students need to develop cultural savvy, which includes knowledge of the tacit norms of the specialism, as well as the range of counter norms (Mitroff, 1974) that operate in scientific disciplines. Together these notions suggest how complex and opportunistic is the process of tacit learning during doctoral study. There are many circumstances in which students perceive there to be conflicting norms that require a carefully thought strategy for action.

A geology student working in an externally funded team-based research area, for example, described his concern about publishing to meet the expected norms of the field: "Here again, this is one of the complexes of the academic community . . . . It's to the advantage of the supervisors to get as many papers out as possible, which is more or less taking a risk with their students . . . . The supervisors push their students to do things before they're ready so that it helps with their funding situation." In experimental psychology, a student described the hidden agenda in her supervision: ". . . and sort of the little games and that, we could have saved ourselves a lot of trouble over . . . . You know, my supervisor tends to play games with people a little bit. Supervision is tied up with the research strategy of the department and there's all this push to get people publishing, to meet your deadlines quicker. But you spend your candidature time . . . on their agendas." A history student described coming to terms with the politics of the research community in a situation where theoretical perspectives were in conflict: "I had a bit of a problem because I had a dispute with one of my supervisors, who was also the chairperson. And this person was very influential in the field. It created a bit of a bad taste, because it sounds like you are whingeing or being some sort of juvenile about things. Eventually a person from philosophy was brought in . . . and it helped to resolve the situation, but it was a question of style, and it was a bit of saving face all round."

Gerholm (1990:263) makes the point that the standing of a student who fails to comply with the implicit rules of the research community will no doubt be affected, and that this may amount to the student forever remaining an outsider to a particular scholarly community. A supervisor in philosophy gave expression to this idea: "I have a student who was determined to do his thing, not much interested actually in philosophical argumentation, but he's interested in ideas. He likes having ideas, but the sort of nitty gritty of testing them, providing arguments, isn't something he wants . . . so the issue is, would he receive favourable responses in terms of enthusiasm for his work or whatever? And by and large, he's had less of that than might have been the case . . . now one is aware, you see, that examiners can read this thing and think, 'who's been doing this, how can they let this sort of stuff through?'"

One important lesson is that it is frequently necessary to defer to the status and authority of the supervisor, mainly because supervisors are gatekeepers in the process of socialisation to a particular academic community. Students who have difficulties during supervision often find that these relate to theoretical perspectives in transdisciplinary or individualistic fields, to struggles or conflicting advice associated with joint supervision, or, as is sometimes the case in experimental science, to not having enough attention from supervisors with heavy externally-funded research commitments. In all such cases, students report that it is critically important to handle the situation diplomatically, in deference to the status and authority of the supervisors in the departmental setting and also in the wider scholarly domain. Sometimes students resigned themselves to their problems; sometimes a workable compromise was found. A physics student, accepting the hierarchical and status-driven nature of his discipline, crystallised this issue: "A student is so dependent on a supervisor, you know. They're influential because of their research. The bottom line is that the supervisor can make or break your career. They can help you get a really good job in the end, or they can set it up so you'll never get a job in a university." In the less hierarchical discipline of history, another student's concerns were similar: ". . . I personally haven't had any problems, but I think a lot of people [doctoral students] don't really know the avenues you have to go through. How to approach journals to get publications or how to get advice from someone really big in the field . . . who do you go to? I mean, because students are so afraid of treading on toes because those people [supervisors] are so important to your program and to how you are taken in the field generally."

4.2. Getting into the Club

Being accepted as part of the scholarly community is the result of behaving in acceptable ways, first within the department and the immediate research setting, then more broadly in the international scholarly community. In doing the work of their doctorates, socialising and benchmarking with others in their research cultures, students assimilate important disciplinary myths and legends, folklore and codes of practice. A professor of behavioural sciences from a highly productive and internationally significant departmental research culture vividly described the benchmarking process: "We socialise them. It's informal, but we all mix outside the department . . . and we run workshops on things. On writing. On what can go wrong, things like that. There's a distinct progression of a group thing being followed in the department. I think we justify what we do [the group thing] on educational grounds, but this is a research culture and an intellectual culture and an investigative culture. And we're open about that and that is largely what attracts the students and keeps them. Even with those students working on their own in very isolated areas, the students get to know one another quite well and they help each other a lot. If something has to be done over the weekend, which it invariably does, then they swap around a lot, and they inform each other, swap ideas . . . there might be three or four of us working on a critical task at a time, so it gets very intense for a couple of weeks. There tends to be a lot of cross communication and assistance and I suppose, therefore, some sense of belonging to a group develops. It's not formal or directive. It just happens because of our proximity and because of similar kinds of research. And the students learn a lot from this in the beginning. They get into the club."

The importance of Gerholm's (1990) classic paper on tacit knowledge in academia is that it articulates the importance of understanding explicit disciplinary norms as well as what we might call the de facto norms of a disciplinary culture. In earlier chapters, a distinction was drawn between the behaviours and practices explicitly expected of doctoral students, and the *savoir-faire* required to fit into the culture and to understand in full measure the tensions associated with conflicting norms. During doctoral study, a clear understanding of explicit disciplinary norms and de facto norms is achieved through benchmarking, a key strategy in tacit learning. The sources for benchmarking consist largely in the members of scholarly communities, from exemplars through to their students, who maintain and generate collective values, traditions and conventions.

A history student from a large research university described the 'club' in her field quite vividly: "There was a time I can remember when I was completely unaware of the politics of the group. Not that my supervisor played a role, really . . . Actually, it was one other student I know, who is finished in this department now. But [she] knew what was what, [she] understood who the Mafia were and who the other ones were. I don't know where she got it from and I kind of resisted listening to it at first. But as time goes on, I find I'm willing to listen . . . I feel I need to know now, it affects how I'm shaping my stuff." Not only had this student realised the importance of getting into the club, but she also had learned important lessons about adapting her own knowledge outputs from the political scenarios she observed.

A student in experimental physics commented on what she was learning during her doctoral studies: "Let's just say that I think we're fairly new players in the game that we want to undertake in our field. We're learning. For example, one good measure of how well known the [research] group is, is how many times your supervisor or head of the group gets asked to referee papers in journals . . . we're not part of the small community which is basically based in America at the moment, . . . [but] . . . in a way, I'm starting to make connections about what works." Here the student expresses an awareness of disciplinary values, behaviours, standards and conventions, arrived at by comparing the performance and practices of her research group with an elite group in the United States. The student tacitly comes to understand and to be able to adopt appropriate norms that are essential to being accepted by that community.

In the social sciences, getting into the club needs to take place at a reasonably early point in candidature, because the development of social savvy is so important to learning outcomes which relate not only to the research culture, but also to substantive knowledge, methods and techniques. Yet it can be fraught with difficulties for the student in fields where, as already noted, there are intellectual cliques to navigate before an academic identity can confidently be forged.

Getting into the club is a different enterprise yet again in the more individualistic fields of the humanities. Here, the imperative is to find out how to assert a personalised perspective in a way that is diplomatic and sensitive to the values of the intended audience. A history student describes the process: "... there are some [students] who are extremely ambitious and instantly try to find little threads of connection with other people and try to climb the greasy pole. I must admit I always forget to do this sort of thing; you have to do it on your own. And I think, 'damn, I should have exploited the situation and I forgot.'"

The contrast between the humanities and the experimental sciences may well be explained by differences in the nature of their knowledge bases. The 'club' for the history student is less well defined; scattered international communities of scholars are unlikely to be a coherent, homogenous group. As a result, that identification with the discipline is not so clear-cut, though of course it is no less important. This notion of there being different degrees of commitment to a field among doctoral students in different settings accords with Whitley (1980:311) and probably accounts to some extent for the lower completion rates and longer completion times in the humanities and social sciences referred to in Chapter 2.

4.3. Checking and Comparing the Learning

In the social setting for research the doctoral student socialises with the supervisor and a range of significant others. In this setting, there are opportunities to bench-mark the values, performance, behaviours, conventions and standards of experienced researchers and peers: in short, to model the means by which knowledge is manufac-tured, and to understand the conventions and traditions governing the reporting of it. The socialisation of doctoral students has been described as a function of what is learned through doing the work of the doctorate, and of being socialised into a com-munity where the tacit rules of disciplinary conduct are played out.

Benchmarking, as described in undergraduate and online learning by Parry and Dunn (2000), is clearly an important element in the development of disciplinary savvy. The novice researcher has to check by comparison with others in the social setting for research on many fronts to ensure that they too develop the skills and capacities, the appropriate disciplinary savvy, that will give them confidence in reporting their new knowledge. That the benchmarking process has, to date, remained unremarked in the related literature is curious, considering the extent to which some researchers have examined the processes of doctoral study. Neverthe-less, it is a means to developing disciplinary savvy, checking and comparing sub-stantive knowledge, techniques, networks, attitudes, approaches and much more. How does benchmarking take place? Reber (1997:154) explained:

In our experiments on implicit learning, our subjects [sic] were conscious of the fact that they have learned something. They are unaware that cognitive change has taken place during the learning phase of the experiment; they know that they know something that they did not know before. They have a "feeling of knowing" and when making decisions or solving problems, confidence ratings correlate with performance.

When benchmarking is effective and disciplinary savvy is developed, how is it man-ifested? Clear evidence of it can be seen in any successful doctoral thesis. We now turn to doctoral theses to examine this evidence in more detail.

CHAPTER 7

LINGUISTIC ACCEPTABILITY

1. THE SOCIAL BASIS OF LANGUAGE

There is a widely held view among those interested in the nature of human experience that different kinds of reasoning are expressed through different forms of language. Originally an idea of Wittgenstein's, Toulmin (1972: 68) has articulated this idea in a way that fits easily with what we know about academic cultures: " . . . the very language through which our enculturation is achieved is itself intelligible only to men who share enough of our own modes of life", he argued. More recently, Becher's (1989: 89) account of the codified and sometimes highly symbolic nature of specialised language in particular disciplines illustrates how specialised disciplinary language is.

Physics, of course, is an easily recognisable example of highly codified language that is largely impenetrable to outsiders. While we can recognise the codified language of contrasting disciplines, it is curious that experienced supervisors are rarely able to verbalise the codified characteristics of their own academic language, although they overwhelmingly report the ability to recognise them in doctoral theses when they see them. This widely shared experience shows how dual processing theories of cognition (see, for example, Epstein, Pacini, Denes-Raj and Heier, 1996; Epstein and Pacini, 1999) may operate. It also underlines the distinction between implicit and explicit learning drawn by Reber (1997), who notes that the empirical evidence for implicit learning is unequivocal. However, it is not well understood how this form of cognition interacts with its more explicit, reflective cognitive counterpart in producing sophisticated learning outcomes. This leaves us with a problem: scholars, including successful thesis writers, do exhibit command of the highly codified language of their disciplines, but how do they do it? For supervisors, there are issues, too. How do doctoral students come to know highly discipline-specific conventions and stylistic traditions comprising discourse savvy when such conventions and traditions are not explicitly taught? How can we help doctoral students to meet the expectations of examiners in terms of disciplinary writing and style?

In this chapter we explore doctoral thesis writing in some detail, identifying in various disciplines a range of field-specific writing characteristics that arise from the epistemological nature of their knowledge bases. The chapter highlights the complex nature of the tacit learning involved and suggests avenues for socialising students into the characteristic styles and conventions of writing in particular fields.

1.1. Knowledge Expressed Through Language

There is a well-established literature demonstrating how specific forms or codifications of language are used as the authoritative vehicle for transmitting the knowledge base of a field, indicating how conventions are regulated and perpetuated (see for example, Bazerman, 1981; Berkenkotter, Gilbert and Mulkay, 1984, Knorr, 1977). Striking contrasts between broad disciplines are drawn by Bazerman (1981), who compared scholarly texts from science, social science and the humanities to show how they reflect phenomena that are not equally distributed across those disciplinary settings: conceptions of phenomena are different, so must be articulated differently. In this vein, Becher (1989: 90) argued, for example, that hard-pure knowledge, such as chemistry, is formulated in an established context, and within a known framework of assumptions. Alternatively, in soft knowledge areas, such as English literature, topics develop individualistically, and so their context often has to be separately elaborated. These features suggest that in order to develop discourse savvy, the elements of social, cultural and grounded savvy, discussed in detail in Chapter 6, all play a vital part.

All of the related research indicates that disciplinary language reflects important epistemological features of the knowledge base it represents, and that the writer must understand these features in order to be able to use them. Indeed, the language features of successful doctoral thesis texts are highly differentiated by their disciplinary contexts (Parry, 1998), which can be ascribed to recognisable aspects of style and structure in reporting new knowledge that vary across different fields of study.

Supervisors from every discipline describe how they feel confident about advising students on thesis writing. Yet they cannot accurately explain how they do this, and most attribute the deficiency to their having had no formal training in linguistics or writing style. Both supervisors and students are operating with a sophisticated 'tribal language' in which the need to become fluent among a specific group of peers is implicitly understood. However, the nature of exactly what is being learned, and of how it is learned, has remained somewhat obscure.

1.2. Conscious and Subconscious Language Learning

In the humanities disciplines, there is a prior assumption of doctoral students' language proficiency, epitomised by a philosophy professor: "We don't give advice about grammar. We expect them to be up to that. But we might make extensive annotations, and we would discuss them with the student . . . We would direct students to other theses, give them an idea that way, but certainly not tell them how to do it. It is difficult. We have certain expectations about structure. I don't know how to describe them to you, but we give students examples and hope they see what it should look like." A linguistics supervisor provided an equally vague explanation: "I would [help the student with the writing] by extensive correction of a series of drafts. We have a style that applies to our work, and we would expect students to stick to that. I don't know exactly what we teach them about their own writing." Comments like these strongly suggest that powerful conventions that are understood by tacit means are underpinned by field-specific knowledge, and that the learning processes employed by doctoral students are sophisticated and largely inexplicit.

Social science supervisors express more willingness than supervisors in other disciplines to help students to abide by the typical writing conventions of their fields. They also appear to recognise the inexplicit character of those conventions.

A sociology supervisor described the tacit nature of the process by which writing style is learned: " . . . I might alter every sentence or every second sentence. If a student is having a great deal of difficulty with one point I might write some sentences and say, 'you know this will in fact say such and such, so you really haven't said what you need'. It's sort of subliminal or it's too subtle, so I'll do those sorts of things." Subtle differences in style and argument structure are especially important in fields such as psychology, where a specialism's values and conventions may resemble science-based disciplines, or where it may operate instead as a social science. Psychologists particularly seem to appreciate the importance of ensuring that students adopt an appropriate approach to structuring their argument in the thesis. An experimental psychologist explained: "Well, I've always told them that in the end the thesis has to tell a story, and that they should always be trying to put everything they can in that context: sort of logical sequences and ideas and findings that in the end interact to make a coherent story. The expression is the most important thing, because, depending upon the field, there is a certain style."

2. UNDERSTANDING THE PURPOSE OF LANGUAGE

Bazerman (1981) argued that the purpose of the argument is a major disciplinary influence affecting structure and style. In science writing, for example, the purpose is to report information. The author assumes that major facts are embedded already in what is reported, such is the accretive nature of knowledge. In social science writing, however, the author must persuade the audience to accept a particular interpretation of the phenomenon being discussed. This form of persuasion is intellectually complex because in the social sciences argument is holistic, but at the same time scholars do not necessarily share methodological or theoretical frameworks. Writing in the humanities is different again: here phenomena must be established as consequential within a personal perspective, and the audience must be persuaded to accept the new insight.

There are three overlapping domains of scholarly writing in the disciplinary groupings of the social sciences, the humanities and experimental science that have been shown to be readily identifiable to doctoral students and to supervisors (Parry, 2000). How these domains work out in written texts varies according to the nature of the discipline and the associated cultural traditions. They govern the typical structure of the argument and its style; the techniques and conventions for citation and appraisal of existing research; and the ways in which writers assert the tacit knowledge of their fields, including the position the writer adopts towards the audience. Each will be considered in detail in subsequent sections. Differences in these three domains are clearly linked to the nature of the knowledge they represent, and they bear witness to the sophisticated nature of linguistic conventions that doctoral students learn to master during candidature.

Many doctoral students expressed concern about how to identify and master the linguistic conventions that constitute acceptable writing at the doctoral level in their particular fields. This concern is not simple: while there are core language features of broad disciplinary areas to which we will turn our attention, scholarly fields are also dynamic and constantly changing. The increase in transdisciplinary and applied doctoral research during the past decade bears witness to this dynamism, as we have seen in Chapter 2. Furthermore, individual specialisms may exhibit counter-characteristics in some fields where generalisations about language features do not ring true. Archaeology within the humanities, economic history within history, or ecology within biological science are notable examples of specialisations with their own counter cultures. For the doctoral student, identification with the knowledge base and its disciplinary area, together with conscious and subconscious recognition of writerly techniques in the field, are critical to the successful writing of the doctoral thesis.

2.1. Different Purposes of Metaphors

Scientific language reflects key characteristics of the knowledge base of science-based fields, where knowledge is concrete, impersonal and value-free. It typically focuses upon material objects and emphasises the reporting of information. Technical terms are used as summaries for sometimes long and complex processes that are represented as entities, creating an abstract language. These abstract technical terms are used to organise the world and create precision in meaning. An example from genetics is *reproductive ecology*. A description of this term in everyday language might be 'the way in which animals produce the next generation, the way that this process affects the animal and its surroundings and the way in which its surroundings affect the animal'. Writers frequently pack a number of processes together as a description of a concrete entity: *peak amplitudes* (amplitudes that have peaked), *short stratigraphically equivalent sections*, *unstimulated rat myotubes* and *fleshy fruiting shrubs* are examples of this kind of package. This technique, where scientific writing emphasises concrete phenomena in a predominantly reporting mode, meets a key expectation in the discipline: providing a non-committal, objective stance with an intended or known audience. In the following text, these features are illustrated, showing how the thesis writer creates objectivity by avoiding direct judgements, although in this case an element of 'maybeness' is introduced through the reference to existing research:

This 'recombinational breakdown' has been demonstrated as a major component of the reduction in F2 and backcross viability between taxa of the grasshopper *Caledia captiva* which differ by a series of pericentric rearrangements (Shaw and Coates 1983; Coates and Shaw in press). The relevance of this mechanism to other types of structural rearrangements (i.e. interchromosomal rearrangements) is not clear. John and Freeman (1975) suggested that Robertsonian fusions will generally lead to a reduction in chiasma unless, of course, recombination was previously restricted to the. . . . (Moritz, 1984: 14)

Here we can clearly see the condensed packaging of science style, where the language is highly technical. It also creates a discernible distance between writer and the audience, suggesting objectivity. The writer seems to know whom his audience consists of, and this understanding permits certain assumptions to be made about the prior knowledge of the audience about the topic. This doctoral student has learned condensed, technical packaging from extensive reading in the field, and from supervisors' and colleagues' emendations.

Sharply contrasting with the characteristics of science are the mode and focus of humanities writing, where the individualistic and interpretive nature of the knowledge base are reflected in the typical features of linguistic style. Language is highly metaphorical and abstract, with an emphasis upon interpretation of phenomena and upon asserting or arguing a particular perspective. Meaning is built through the use of metaphors that condense quite complex, and sometimes esoteric, ideas into concepts. Terms taken from history, music and Australian literature, such as *the rhetoric of violent struggle, a narrativised preoccupation, popular culture, mimetic principle, majority appeal*, all capture complex concepts in small packages of metaphorical language that are highly visual and evocative in their appeal, though such terms are rarely technical. Metaphorical elements of language also give a particular interpretation of the phenomena; for example, *commercialism in popular culture*, taken in its context, suggests a particular perspective of the author. A philosophy supervisor remarked that techniques such as this help to "de-ambiguify" the writer's intention in the argument. The following text from a doctoral thesis in music illustrates how inherent ambiguity is reduced and qualified:

At the heart of these arguments lies a lack of support for the Australian artistic product which has traditionally come from the ABC [Australian Broadcasting commission} but which is now being undermined by the power of commercialism in popular culture. Given the continued homage paid to art music from elsewhere – the consequence of a highly impressionable society – the absolute necessity for governments to win majority appeal, the suggestion that "This generation of politicians is the last that will have any deep affection for the ABC" is a reasonable one (11). It is a suggestion given validity in the erosion of the ABC's funding in recent years and the proposed cut of $18.7 million for 1988–9 (12). In effect, the erosion of funds is an erosion of the ABC's role as outlined in its charter. (Morton, 1990: 36)

Here the student densely packs the argument with metaphors: those such as *the power of commercialism*, and *a highly impressionable society* are amplified by the argument, which is clearly enunciated in the first sentence – where the nature of the topic idea is conveyed – and then qualified in the last sentence. A different piece of text, from a doctoral thesis in linguistics, exhibits similar characteristics:

If, as generally agreed, what a speaker can mean in language is the primary measure of his language ability, the investigation of the meanings expressed becomes central to the description of any corpus of language. To investigate meaning, and at the interface between meaning and form, to interpret semantic distinctions in the utterances of children, it is necessary to have a knowledge of the context of those utterances. But context is itself a relational notion, which must be defined, in relational terms. What is needed is an operational framework within which to investigate the meanings encoded in communicative language and that framework must include the relationships being postulated between areas of meaning and areas of context. The major constituents of such a framework are suggested in Halliday's statement that speech is 'typically relatable to its context of situation in recognisable and systematic ways . . . ' The meanings reflect the field, tenor and mode of the situation. (Walker, 1980: 11)

Each of these texts illustrates how writing in the humanities relies on the use of argument rather than on a reporting mode of writing. The writers' purpose is to assert a new perspective on existing phenomena, and in these examples they show that they know how to use a store of relevant material to provide a persuasive context for their perspective. The effect is highly personalised and highly subjective argument that depends upon establishing credibility with the audience.

Purpose and associated language features in social science fields are different again. Knowledge is concerned with constructed ways of perceiving human phenomena. Social science language tends to be metaphorical and abstract, reflecting the individualistic and interpretive nature of knowledge, yet it is also technical (in some fields more so than in others), and there is a strong emphasis upon causal relationships between concepts. These features reflect the importance in the social sciences of establishing an appropriate theoretical and methodological framework that the audience can accept as the justification or basis for asserting an argument. Contrasting examples from economics and education show how students combine these elements to make specific, theory-laden meaning that is interpretive:

Beamish (1984) in his study of 66 co-operative agreements focuses on mutual long term need between partners as an important issue in assessing the agreements' potential. He comments that cooperative agreements are often formed because of uncertainty. Apparently learning follows the agreement and then the need for the partner wanes. Therefore, there is value in finding a partner with the same short-term and long-term goals. (Berg, 1992: 75)

Although teachers new to the school questioned the sectional interests of the clientele, cultural tensions, as illustrated in Chapter 6, were cast in terms of teacher conception of leader as culture builder developed in the literature . . . (Proudford, 1992: 203)

In both these examples, the mode of argument is explanatory but the argument is asserted from within a framework that has been established earlier. The texts feature a high level of metaphorical language. Depending upon the field or subject, there may also be a high level of technicality. Concepts that would be unwieldy to explain in common parlance are grouped together, or abstracted, using metaphors such as *mutual long term need*, *sectional interests*, *culture builder*. These thesis writers have developed a sophisticated use of metaphor based upon the terminology of their fields and upon a tacit understanding of those metaphors that are likely to be understood, and accepted by the audience, as against those that are not. In commerce, for example, a supervisor explained how students develop these skills: " . . . we have a training ground . . . they get a lot of practice in honours year, and we make corrections and suggestions till they learn the concepts and how to put them together. You're giving them what is current today, so they have to read to be up to date with the jargon and what it means and how you use it. You hope they're going to add to it, so you have to keep showing them where the frontiers are, teaching them to be critical".

Comparing the writing of the three disciplinary areas, the least transparent to the layperson is scientific writing: in some specialisms, such as theoretical physics, the highly technical language may be quite impenetrable to an outsider. This characteristic, argues Whitley (1984), gives rise to a highly formalised symbolic language in the field, with a standardised and formalised reporting structure that transcends geographical and social boundaries. In contrast, writing in humanities specialisms is more penetrable, reading more like everyday spoken language, providing detail and explanation.

However, in the social sciences, which exhibit individualistic and ambiguous messages that are open to interpretation, style is governed by the message being communicated, the level of abstraction and the extent to which colloquialisms and universal terms are used. In fields where there is a strong interrelationship with the laity, the language is likely to be more accessible to outsiders.

Individual students for a doctorate do not usually understand writing conventions in their specific fields on a conscious level. However, doctoral study is geared to provide students with effective socialisation into the writing culture of their fields: there are thesis models, opportunities for writing practice and advice through various reading and writing tasks, including drafting the thesis itself, and of course there is easy access to the writing of the exemplars in the field. Students whose supervisors are active writers within their specialism are at an advantage; those whose supervisors are active in divergent fields usually need to seek for themselves opportunities for learning the subtleties of disciplinary writing style.

3. STRUCTURE OF ARGUMENT

From the student perspective, many conventions for structuring an argument are implicit, but because they fit so closely to the epistemological features of the discipline, they make clear sense and are learned easily. These features exist at several levels: there is the level of the overall structure of the thesis, and then, within that, there is the structure of argument in sections or whole chapters, which serve particular purposes, such as to report, discuss, explain and justify knowledge being claimed. The sections have different purposes, which together constitute a coherent whole – a totality that, according to the nature of the knowledge, asserts a particular attitude towards its audience. Students come to know how to construct their argument according to the purpose of the section of writing: "I'd have something I can conceptualise but when my supervisor looks at it, there's something not right. So he'll write in a question or a comment, and when I see it, I think, 'Oh, yes, of course', and then I know . . . subliminally" (economics student). The 'right way' seems to fit with the student's prior knowledge of how argument is constructed, because the student recognises the characteristics of the knowledge base in the writing. Becher (1987b: 56) observes how this kind of recognition works in scholarly publications:

Pure science, as exemplified in physics, is based on a process of accretion of knowledge, and hence may be said to be atomistic (each new piece of the jigsaw has to be identified and fitted in to what is already there). History and sociology, being in their different ways based on criticism, reiteration and reinterpretation, do not match the jigsaw metaphor. Knowledge here is organic and holistic, growing in a complex and comparatively unpredictable way.

3.1. Argument in Science

In science, the writer knows the rules and components of reporting and can build the argument from 'the bottom up'. This fits with the concerns frequently expressed by science students, here represented by an experimental physics student nearing completion: "When you are writing up, . . . you may not know yourself when you have got enough to submit . . . you just keep putting it together". The 'bottom-up' form of argument in science is reflected in the typical presentation of doctoral thesis sections. The thesis is broken down into smaller parts that together comprise the whole. These may be reported as individual topics with their own methodology, results and discussion sections, or they may be reported as different experiments or sections within overarching chapters containing methodology, results and discussion. In an extension of this latter form, it is permissible in a growing number of scientific fields to submit sets of published articles for the doctoral degree: this practice is a clear expression of the piecemeal nature of research where a problem can be divided into smaller parts.

3.2. Argument in the Humanities

In contrast, writing in the humanities is more interpretive, not being based on a standard set of assumptions; paradigms, if they can be called such, are personal and individualistic. Causal connections demand complex forms of reasoning in which judgement and persuasion must be strongly asserted. Explanations therefore have to be constructed in relation to the whole perspective, and so argued from the 'top down'. This kind of construction also reflects the patterns of topic determination and development reported in humanities fields, where resolution of the exact research question may take considerable time, but where the resolution itself equates with having determined a vision of the overall argument. Development from the 'top down' is a process in which the writing is undertaken when a vision of the coherent argument has been predicated. In the process of developing the vision, the writer resolves the interwoven issues of method, evidence, materials, structure and development of argument, and so on. This process of constructing the argument after it is fully conceived, as in the humanities, is described by Bazerman (1981: 378) as one whereby codification and the use of literatures in support of the writer's vision are constructed idiosyncratically. A history student described this strategy and how it meets audience expectations in the discipline: "How do you know whether you're an idiot? I guess that I didn't . . . until I had sorted all those fiddly details. . . . And then I had this overall picture. Then I felt confident and started writing. As I go, my supervisor makes these subtle, very slight suggestions that help her to understand my argument better. I feel I know whether she's right or wrong because I know what I'm trying to say."

Argument structures in the humanities, as outlined in typical tables of contents in theses, exhibit a certain inherent logic or 'flow'. The logical sequencing of the thesis is consistent with the construction of the argument from the 'top down', even though the particularistic nature of research in humanities fields gives rise to unlimited diversity of style and form.

Typically, the flow of overall argument in doctoral theses in the humanities begins with an introduction, moves to explain the focus and importance of the topic, then the nature of the problem and gaps in existing interpretations of the phenomena, including the boundaries of the thesis. This introductory, explanatory strategy permits the audience to identify with the writer and to develop an affinity with the writer's perspective, which may be especially unusual or different from existing knowledge. The body of the thesis, setting out the content of the argument, is developed in an idiosyncratic way. In the table of contents, this idiosyncratic argument is usually represented by a set of signposts for the chronology or narrative that it represents.

Finally, there follow the conclusions. In this kind of logical progression, sections of the argument cannot stand alone. The argument structure is developed holistically, according to a particular vision, even though the treatment is individualistic and there is no standard format for reporting. A humanities student explained: "I wanted to make an unusual statement. It went against all the accepted methods, but I was doing it anyway. I guess the organising feature of my [mother's] story was a set of concepts. . . . After a while the pattern falls into place, and you can write it in steps because you know what you're on about."

3.3. Argument in the Social Sciences

A different set of characteristics in the knowledge base of social science conditions the nature of argument and the corresponding structures of doctoral theses. Cast somewhere between the sciences and the humanities, social science theses represent widely divergent fields within the disciplinary grouping. The fields range from ones whose patterns of knowledge production and communication resemble those in science to ones that exhibit characteristics prevalent in the humanities. The structures of doctoral theses, as illustrated by their tables of contents, reflect these wide variations, so it is difficult to make generalisations about structure and style. Notwithstanding this, there are certainly field-specific conventions and implicit rules that doctoral students in social science fields must master in order to write appropriately to a specific audience. Typically, students examine a range of successful theses in their own or related fields and choose an overall format that suits their frame of reference, methodology and style of reporting.

While styles and formats vary widely, there are strategies in the social sciences that students must master, and many supervisors report giving students a range of reading and critical reviewing tasks as a way of modelling these strategies. One important and consistent feature of the structure of social science argument is its interpretive nature, yet its success is also heavily dependent upon the establishment of a viable theoretical and methodological framework that justifies the interpretation. As Bazerman (1981: 378) explains, argument in the social sciences must establish that a phenomenon exists and is of consequence. It must reconstruct existing literature to create a framework of thought and criteria of proof with which the audience can identify, in order to claim some certainty in its conclusions. An examination of a small piece of text shows how a doctoral student manipulates these characteristics:

Risk is generally discussed in the context of environmental risk. The concern is the exposure of assets that will be affected by changes in political, economic, financial or competitive conditions. To limit the exposure one firm may face, an interfirm cooperative agreement is considered. Avoiding risk becomes a major motive for joint venture when managers pursue the implications of portfolio theory. Porter (1987) suggests that the concept of corporate strategy most in use is portfolio management and defines it loosely as diversification through acquisition. He goes on to say, however, that portfolio management is no way to conduct corporate strategy unless you are in a developing country. Portfolio management becomes particularly important in exploratory activity, when distribution of returns is highly skewed. It is necessary for firms engaged in such activities to have a large portfolio of projects to achieve diversification benefits (Berg, Duncan, Friedman, 1982). Interfirm co-operation facilitates the acquisition of a larger portfolio with reduced risks. (Berg, 1992: 52)

Here the writer develops an argument that relies heavily upon existing theory, yet it advances a new, subjective interpretation about the reduction of risk in interfirm cooperation that the audience can accept as plausible. The confidence with which the conclusion is reached is dependent upon the framework of thought having been carefully constructed to provide a particular, credible direction in the argument. The language is metaphorical, as in the humanities, but it is also technical, as in the sciences. The packing together of complex ideas is achieved through the use of metaphor (*environmental risk, interim cooperation*) and technicality (*portfolio theory, distribution of returns*), though in this particular example owing to the nature of the field of study there is not a strong reliance upon technical terms.

4. LINKING IDEAS

The overall structure of argument directs, among other things, the mechanisms by which ideas are linked in doctoral theses. In science, standardising the reporting procedures reduces ambiguity. Swales (1983: 192–3), in an analysis of the structure of writing found in academic journal article introductions, describes how formalised reporting structures work in practice. Writers use a set of moves to cap an existing research story with their new research story, working from establishing the field to narrowing the field and preparing for the present research, then identifying a problem or gap in the existing literature, then explaining how the present research will address the problem or gap. Such a set of moves provides a framework for the argument and permits ideas to be linked in a way that audiences find familiar.

4.1. Linking Ideas in Science

The following example of text, from marine biology, illustrates how a writer builds a platform for her doctoral research by identifying an unresolved problem:

Competition amongst coral species has received considerable attention, with the result that its existence is generally unquestioned, but its significance as a factor controlling coral distribution and abundance, and hence community structure, has been the subject of some controversy (Sheppard, 1979, 1981, 1982; Bradbury and Young, 1981b; Cope, 1981, Bak et al. 1982).

Despite these studies of many aspects of coral ecology, only a few authors have attempted to integrate data on the population ecology of coral species with a study of the contribution of these attributes to community structure . . .

[lists relevant literature].

Perhaps the area of greatest interest to coral ecologists over the last 5 years has been the process of maintenance of high diversity, in particular the roles of disturbance and succession in this process . . . Theories on diversity maintenance applicable to rainforests and coral reefs were reviewed by Connel in 1978. Models of diversity maintenance via disturbance include the 'intermediate disturbance' hypothesis . . . A problem in comparative work is the determination of how 'high' diversity is and how 'intermediate' disturbance is.

In this study, the distribution, abundance and diversity patterns of a scleractinian community in a patch reef environment at Lizard Island . . . [are] investigated. Aspects of the population ecology of some component species in the community (reproductive ecology, settlement, recruitment and mortality) [are] studied, and the interactions between coral ecology and community structure . . . assessed. (Harriott, 1983: 3–5)

The writer argues that research interest in relation to the problem has been in a particular direction (*diversity and disturbance as factors in population ecology*), then caps the existing research story by offering a new theory from a different field altogether, which ought to be applied (*theory of diversity maintenance via disturbance in rainforests*). Then, in the last paragraph, she reinforces her story by stating how the theory will be practically applied to the topic of *competition amongst coral species* heralded in the opening sentence.

The above text also illustrates another typical characteristic of the standardised nature of science writing. The linking of paragraphs within sections tends to be organised according to the "theme and new" method described by Halliday (1994), in which the first sentence of a paragraph is an introductory or topic sentence. The remainder of the sentences in the paragraph then contain new information about the topic sentence. The overall effect is that the topic sentences convey the main ideas that are the outline for reporting the new knowledge. These structural features work to produce a disciplinary language that meets audience expectations by conforming to stylistic norms. The effect of meeting audience expectations in the reporting of knowledge is that significant epistemological characteristics of the knowledge base in science, such as objectivity and a focus on evidence, are reinforced. While few science supervisors can articulate this pattern, they certainly recognise it, and their emendations on thesis drafts reflect their conformity to it.

4.2. Linking Ideas in the Humanities

In the humanities, the overall structure of argument and the writers' strategies for achieving cohesion do not exhibit a standardised format, as befits the particularistic nature of these disciplines. For the student, the wide variations possible in thesis format can be unsettling. While in science the structure of argument is built upon accumulated material, enclosing the new knowledge within a framework of existing knowledge, writing in the humanities is not fixed in a knowledge framework. Instead, it offers a perspective that illuminates the phenomenon itself. The means by which this is achieved in humanities fields is individualistic, so it does not fit with the reporting procedures of fields where knowledge is cumulative. For the student, working from 'the top down' is imperative.

Here too, as in other disciplines, there are particular strategies for the student to master, usually through developing expertise with typical texts. In the body of the thesis, as in other forms of writing in the humanities, a narrative that is the framework for the argument conveys logical progression. The narrative is not just a story, however. It must meet certain disciplinary requirements, as Bazerman (1981) has observed: the author must establish that existing perspectives are inadequate, and persuade the reader that the new vision is not only relevant but also adds greater insight into the phenomena than was possible previously. How a narrative provides a framework for the argument can be seen in this example from the introduction to a thesis concerning Australian literature:

The self-reflexive preoccupation of non-realist writers with beginnings (and conclusions) is a preoccupation with a search for origins, with the signifier in search of its signified, language in search of its meaning, identity in search of its reflection. It is a narrativized preoccupation playing upon and foregrounding those repressions which enable the realist text to disguise the desire-ridden and thus compromised nature of its procedure in order to close its search. The self-conscious beginning, however, is not always mounded in the same form nor does it always achieve the same ends. 'In the beginning was the Word and the Word was with God, and the Word was God'. Umberto Eco opens the Prologue to *The Name of the Rose* with this quotation from the supreme Christian authority: the word, as original object, becomes an attachment and possession ('*with* God') and then an embodiment of its owner ('*was* God'). The adventure of the word(s) in this novel is going to be an adventure taking place within the already created labyrinth where its parodic echoes can ring out. Kate Grenville's *Joan Makes History* contests the authoritative beginning quite differently: 'In the beginning was nothing much.' Thus begins this novel which overwrites the history of Australian settlement with the voices of some of those female groups who were previously voiceless in so far as the official, patriarchal, white records are concerned: Joan the convict, washerwoman, the government official's wife, the Aboriginal woman, the mother. . . . (Gillett, 1990: 3–4).

Here a body of significant and interrelated detail is given logical cohesion by the narrative, which is the writer's own creation and intention. In this text, other features of humanities writing, such as longer, explanatory clauses and the use of unacknowledged metaphors from the literature as though they are universally understood (*signifier, signified*), are evident. The emphasis in the style is upon the author's interpretation of the phenomena, and the use of dense metaphors has the effect of packing complex ideas into manageable words or phrases that are conceptually rich, visually evocative and intellectually inviting.

Paragraphs and sentence structures tend to be longer and contain more explanatory clauses in the linking of ideas in the humanities than in other disciplines. The logic extends from one sentence to the next, rather than adopting the 'topic and new' information structure prevalent in science. The linking of ideas from one sentence to the next provides a structure for the argument through the development of the narrative. This text, from a thesis on music, illustrates the importance of narrative in linking ideas in humanities fields:

While it has maintained a monopoly in setting Melbourne's orchestral standards the ease with which the best players were once attracted to the pioneer orchestra no longer exists. Orchestral resources have been diversified by the growth, complexity (including a wider cross-section of nationalities) and competitiveness of an artistic environment which has become increasingly sophisticated. Social and economic conditions have changed the role of the orchestra in society and consequently the status and conditions of the professional musician. Since the 1930s the ABC's role in continuing its maintenance of the orchestra has created a relatively secure environment for its musicians and in turn has shaped much of its character and identity; an identity aligned with the conservatism long associated with the ABC. This perceived conservatism emanates from the non-commercial, non-competitive nature of the organisation. Indeed, the term 'Auntie' has for years been applied satirically to the ABC in describing its attitude towards various matters of public interest. In recent years it has been subject to stronger, more cynical comment. However, the view that with its "orchestras, its' educated' voices, and its associations with various establishments such as opera, ballet . . . it has seen itself as having a 'civilising' role, promoting essentially bourgeois (and British) cultural practices and values . . . " should be seen in the context of Australian society and culture. . . . Australia is a country with a shallow musical tradition, and values which do not include art, music as a fundamental component of the culture. Seen in this perspective, the ABC plays a unique and vital role, the extent of which can be gauged by examining briefly its charter and its current situation. (Gould, 1989: 31–2)

This example of text tells a special story. The writer begins with a topic sentence that needs sophisticated qualification. The author's perspective, reflecting a personal stance, illuminates the phenomena under view.

4.3. Linking Ideas in the Social Sciences

The strategies are different again for doctoral students in the social sciences, not least because research topics may fit anywhere along the disciplinary spectrum, as we noted earlier. Like humanities writing, argument in the social sciences asserts a personal interpretation, though the interpretation is more obviously theory dependent. Because there are co-existing as well as competing paradigms, there is no uniform framework of thought, so writers must establish a framework for understanding with the audience. Bazerman (1981: 378) observes that the "sociological literature on scientific behaviour is more diverse, unsettled and open to interpretation; therefore, the essay must reconstruct the literature to establish a framework for discussion." The objective of the social science writer, he argues, is to "urge, persuade and direct" the audience along the lines of the author's thoughts. While Swales' (1983) story-capping strategy works in fields where knowledge is cumulative, Becher (1987b: 57–8) points out that it does not work for those in interpretive disciplines:

[story-capping] . . . clearly depends on a field in which knowledge either is, or is presented as being cumulative. It underlines the need for academic authors in such fields to establish the credentials of their work in terms of what has gone before – a necessity which historians are spared by the very nature of their task, and which sociologists can avoid unless they wish to adopt a scientific stance.

That is to say, theses in social science fields may exhibit features closely resembling those of science-based fields, or they may be more reiterative and interpretive, as in the humanities. For students, there is a need to establish a single disciplinary focus and style of argument, and fully to understand where their doctoral research fits on the disciplinary continuum.

There are techniques to master within the overall framework, too. Sections of writing in doctoral theses in the social sciences, consistent with the objective of establishing a common framework of thought with the audience, have a range of purposes such as reporting, explaining, justifying, describing and arguing, though the dominant mode is explanation. This emphasis in thesis writing has the effect of giving subjective interpretations an almost objective appearance by meeting the strong justificatory requirements of social science disciplines. The following example of text, from a doctoral thesis in education studies, shows how the author relies upon existing literature to provide a framework for understanding the phenomena under review:

Studies of students from disadvantaged family situations, and the effect that disadvantage may have on those particular students, have been topical for several decades. Interestingly, only a limited number of studies has focused on students from single-parent households and school retention but the consistent finding has been that a link does exist between parental absence and early school termination (Ekstrom et al., 1986; Motsinger, 1993; Rumberger, 1983). Other researchers have been concerned with the impact that additional secondary schooling, involving time away from the family, has on the family unit. Boylan and Braggett (1988) argued that, for some students, participation in post-compulsory school situations (eg. boarding school) can cause disruption and dislocation within rural households especially, thus leading to premature senior school termination. This line of thought has been echoed by Cawthron and Craig (1980) who found that parental pressure to leave high school early was stronger in rural families than in metropolitan families. They surmised that this situation existed because rural parents were disenchanted with additional schooling and probably wanted their children to assist with farming duties or to draw another income. (Hemmings, 1994: 29)

Linking from one sentence to the next develops the cohesion in this text. The argument about the effects of family disadvantage on student continuation in secondary school is developed mainly through successive explanation and elaboration and in this way reflects a particularistic perspective. To support the perspective, there is a heavy reliance on existing theory.

4.4. Conventions for Citation and Critique

Conventions for citation in scholarly writing reflect the nature of the phenomena and traditions for referring to existing research in particular fields. As Becher (1987b: 49) explains, "the terms commonly used for the purposes of approbation and disapprobation help to pinpoint disciplinary values and to mark the defining characteristics of any given field." Mastery of the conventions for citation and critique is, therefore, an essential element in demonstrating disciplinary knowledge, particularly in view of Gilbert's (1977) account of referencing in scholarly writing as a covert form of persuasion. The attachment of disciplinary meaning to such conventions implies that, in a doctoral thesis, appropriate citation and critique signals the espousal of appropriate values, etiquette, style and cultural savvy.

Swales (1983) describes different approaches to citation in scholarly journals as being either strong or weak author orientations or subject orientations, though he does not link these to differences in epistemology or disciplinary tradition. For doctoral students, however, there are implicit disciplinary rules about citation to be learned.

4.5. Praising and Blaming in Science

Consistent with the characteristics of the knowledge base of science, where knowledge is accretive and concrete, and where research is undertaken within an existing and well-established paradigm, citation practices emphasise the reporting of information. There is a strong focus upon building a platform of existing research through the judicious assembly of appropriate and relevant material. This kind of platform enables the writer then to assert new knowledge against a background of established knowledge, thus adding to existing information rather than seeking to persuade the reader about a new perspective, as would be the case in the humanities.

In science, citation emphasises material objects and processes, and authorship is usually acknowledged in parentheses, in some cases at the end of information-rich sentences. The following example, taken from a doctoral thesis in genetics, illustrates these citation features:

It has recently become apparent that the chromosomal mutation rate may vary in time and space within a species. With the rediscovery of transposable elements several authors have suggested that they may at least partly explain some examples of rapid chromosomal change (Baker and Bickham 1980; Bush 1981; Patton and Sherwood 1983; Larson *et al.* in press). Indirect evidence for this process has been obtained from cytological studies of hybrids between incipient species (Shaw *et al.* 1983) or between allopatric conspecific populations (Peters 1982) which demonstrated increased chromosomal mutation rates. The evolutionary role of transposons must be considered with caution since our knowledge of the control and frequency of transposition under natural conditions is fragmentary. Although there is a strong case for their mobilisation by genomic stress (McClintock 1978; Shaw *et al.* 1983) there is no evidence that environmental stress will cause their expression as McDonald (1983) envisaged. The relevance of transposable elements to chromosomal change in the absence of genomic stress (eg. hybridisation) remains to be demonstrated. (Moritz, 1984: 23)

Here there is a strong information orientation, with authors being acknowledged usually after their findings have been reported. Towards the end of the paragraph, the information about *mobilisation by genomic stress* assumes more importance in the argument, and the authors of this evidence are acknowledged within the sentence, emphasising the case these authors have made to the argument at hand. Even when there is a need to evoke a particular author within the sentence in order to emphasise the influence of a particular scholar's research, there is still a strong information focus, as illustrated by the following text, from animal biology:

Accurate estimates of the energy requirements of the calling behaviour of frogs have been made, and these suggest that calling is very energetically demanding and the conversion of metabolic energy to acoustic energy is very inefficient. (Ryan 1988). Brackenbury (1977) estimated the energy efficiency of crowing of the domestic fowl (*Gallus domesticus*) by indirect methods and also found it to be energetically inefficient. Read and Weary (1992) also considered the energetic cost of singing and found that song output appeared to be energy limited in species which had higher metabolic rates than would be predicted from their body size, and these species had lower song rates than those with 'normal' metabolic rates. Across bird families, resting metabolic rate (RMR) is very strongly correlated with body weight (Bennett & Harvey 1987), and this also holds true within estrildines (Marshall and Pritzinger 1991). Bennett & Harvey (1987) found that across 78 families, 97.5% of the variation in RMR could be accounted for by body size, and this fitted the following equation . . . " (Dunn, 1993: 51)

There is no overt judgement of previous research other than by inclusion or omission. Doctoral students in science learn this form of reporting as a matter of course, and doctoral programs typically provide a raft of opportunities for students to do so.

4.6. Praising and Blaming in Social Science

Thesis writers in social science fields have to adopt an approach to citing existing literature that may include, as part of the sentence, reference to the author(s) of the ideas raised. Here, because knowledge is diverse and dynamic, the writer must establish a common conceptual ground with the audience, which means identifying with the perspectives of chosen researchers. In fields such as politics and sociology, where paradigms co-exist and intellectual trends and cliques are cultural features, it is important for the doctoral student to signal to the audience exactly which values are being espoused and which researchers are being identified with. Citation practices vary between emphasising the information in the literature, as in science fields, and emphasising authorship, which is the more usual practice. Authors may be cited as part of the sentence or they may be referred to in the argument to evoke a conceptual stance. The following example of text from economics illustrates the author orientation and conceptual stance:

. . . Harrigan's 1988 study found that a larger percentage of IFC were formed by partners who were horizontally related to each other. This led her to suspect that cultural homogeneity among partners is more important to IFC success than symmetry in their national origins. She also found that most ventures formed in recent years were between partners with dissimilar venturing experience levels.

If we accept, as Gray and Yan (1990) do, that a key issue facing managers of IFC is the dynamic changes in bargaining power and industry characteristics, and that these have to be accommodated, plus

IFC structure has to be adjusted to enhance performance, then it follows that the negotiation skills of the partners in securing competitive advantage for their firm is paramount. Experience in previous IFC may develop those negotiation skills, and compatibility of partners builds the trust conducive to successful negotiations (Harrigan 1988). (Berg, 1992: 75)

Here the writer links important aspects of *Harrigan's* research by emphasising *Harrigan* as the originator of a kind of research; this emphasis the influence of particular authors is picked up in the reference to *Gray and Yan*, whose reputations are called upon to add weight to the development of the argument.

4.7. Praising and Blaming in the Humanities

The approach is similar for students in the reiterative, particularistic fields of the humanities, though the acknowledgments tend to be less frequent in the body of text, so distractions from the narrative are reduced. Acknowledgments may be focused on the source of the research, or on the information contained in the research, depending on the field and the purpose of the section of writing. For example, in history, the source and the perspective being acknowledged are inextricable, so that the source becomes integral in the writer's argument; additional information about the source is footnoted, as the following example from a doctoral thesis attests:

. . . R. L. Brett and John Kenyon respectively found Parker's style so 'infinitely tedious', so 'solemn and stodgy' that, they imply, only the most resolutely learned man could have endured reading them.[38] John Coolidge suggests that Marvell's movement between jest and earnest was informed by the need to satisfy two disparate, but educated audiences: court wits who surrounded the king and supported his moves towards indulgence, as well an nonconformists who might have disapproved of the use of jest in a debate on religious issues.[39] . . . .

[38] R. L. Brett, 'Andrew Marvell: Life and Times', in R. L. Brett, (ed.) *Andrew Marvell: Essays on the Tercentenary of his Death*, published for the University of Hull by the Oxford University Press, Oxford, 1979, p23. John Kenyon, 'Andrew Marvell: Life and Times', in J. Dixon-Hunt, *Andrew Marvell*, p. 169.

[39] John S. Coolidge, *'Martin Marprelate, Marvell, and Decorum Personae as Satirical Theme'*, p. 530. According to J. H. Wilson the label wit was attached only to one who made some real pretence to distinction as a poet, critic, translator, raconteur or man of learning. The great body of English people hardly knew that the wits existed except some of the more scandalous exploits were reported with lively exaggeration. *The Court Wits of the Restoration*, London, Frank Cass, 1967, p. 24. (Leslie, 1993: 102)

Here the writer maintains a line of argument that is reinforced by established perspectives, and additional information about the writer's stance is footnoted in order that the narrative is sustained. An example from linguistics provides some degree of contrast in that the view of established researchers is directly quoted to reinforce the argument of the writer at hand:

. . . There is evidence that, by and large, little of the information available to this time about the language of school-age children has been found by teachers to be of direct practical value. Raban and Wells (1979) report that their graduate teacher-students had "very wisely, not attempted to absorb and apply linguistic theory that they had met in their courses, for they had not found it to be particularly relevant." They also report that teachers found people such as Ken and Yetta Goodman, Frank Smith, and Courtney Cazden particularly helpful, because they perceived them to be less linguists or psycholinguists than educators with aims similar to their own. Cazden (1976) postulated that "If theoretical knowledge is to be helpful, it must be carefully selected and restructured into an action-oriented programme. (Walker, 1980: 4)

Here the writer builds an argument in part upon those crafted by previous researchers.

4.8. Citation and Positioning

In order to achieve linguistic acceptability in their fields, as indicated by positive judgements from examiners, it is important that doctoral students learn appropriate but subtle techniques such those described above. While reading in a field and writing practice with scholarly appraisal are two avenues for gaining this knowledge, there is an inherent difficulty in that the techniques may be very subtle or altogether implicit. It must also be said that some conventions cannot be made concrete, for they may be as subtle as a nuance, or a gesture that conveys a certain depth of meaning that cannot or ought not be verbalised. However, students report high regard for supervisors who openly discuss field-specific language with their doctoral students as a regular part of the supervision process.

Students need to demonstrate that they know the etiquette of communication of knowledge among scholars in their fields because it signals acceptance of disciplinary values. In science, or science-oriented writing, an impression of objectivity and of making a contribution to a cumulative body of knowledge, is reflected in typical forms of acknowledgment. In more interpretive fields, it is important to signal whose perspectives are being evoked as part of the argument, because doing so aligns the writer with particular schools of thought and values in scholarship.

The etiquette with which doctoral students make reference to existing literature is another important element of disciplinary lore. Becher (1987b: 52) calls this "praising and blaming", and as the terms suggest, the practices mirror the dominant values of particular fields in quite subtle ways. Doctoral students certainly cannot afford to make offensive judgements about their senior colleagues, whose approval may be sought in the examination process, so it is especially important that they abide by the unspoken rules of their specialisms.

In the examples of text provided earlier in this chapter, it can be seen that in fields such as animal biology and genetics, praising and blaming are done largely by inclusion or exclusion, rather than by making overt judgements. Relevant research is judiciously assembled, in a fairly linear way, to build a platform for the writer's new claims The process is accretive, so the writer is on safe ground with the audience: the level of risk in making new knowledge claims should not be too great. In the case of reporting inadequate or inconclusive research, as in the Moritz thesis text quoted above in referring to existing work by *McDonald*, the criticism is extremely circumspect. The inconclusiveness of findings is identified without making pointed criticisms.

In more reiterative fields where the structures of knowledge are less clear-cut and argument is interpretive, students must learn to make more overt judgements about existing research. As a supervisor in philosophy remarked: "Well, in our field [logic], you have to knock somebody else off to put your own views forward. That means that sometimes you have to be quite bold, and this is sometimes quite hard for students to do, since they are students, after all." The above examples from humanities theses illustrate strong judgements being reported in order to assert a new perspective. Criticism is made on the basis of the craft of the writer in expressing an individual perspective, an inherent part of asserting one's argument in humanities disciplines, in sharp contrast with science.

In social science fields, where knowledge is dynamic and there are competing paradigms, overt criticism of existing research is part of establishing and arguing a case. Here, scholarly criticism can be even more caustic than in humanities fields, with the focus of disapprobation frequently attributed to matters of method. The general absence of such caustic comment in theses is understandable. While it may be necessary to "knock off" another's perspective in order to assert one's own, doctoral students can hardly afford to ignore the values and backgrounds of likely examiners, especially considering the increased likelihood of conflicting examiners' reports in these fields (Parry & Hayden, 1994: 41).

A successful doctoral thesis depends upon the writer managing these conflicting norms. In fact, in Gerholm's (1990) terms, this kind of hurdle is a fundamental part of demonstrating disciplinary expertise. And yet, these conflicting norms appear to be on the increase, as the nature of research becomes demonstrably more applied and transdisciplinary. Oddly enough, humanities and social science supervisors who expressed a view at all on the issue believed that students simply imitated practices in the literature. In only a few cases did supervisors provide explicit advice to students about citation and acknowledgment, which are apparently regarded, as one supervisor put it, as "undercover issues".

It is not surprising, therefore, to find that a widely held concern among supervisors is the degree to which their students should be assisted with thesis writing. While these concerns may derive from the lack of a contract, and of clearly specified expectations by both the supervisor and the student at the beginning of candidature, they may also reflect the lack of opportunities to learn and practice the traditions and conventions for writing in the field and how to differentiate among them. As research topics become more transdisciplinary and idiosyncratic, the need for affirmative discussion between supervisors and doctoral students increases, to say nothing of the need for students who are writing in other than their first language. It is worth considering at the start of candidature the supervisor's and the student's writing ability in the field, the external supports available to the student for advice about writing, and what kinds of models and opportunities to practice using the language features of the field can be made available. At the heart of this issue is an institution's capacity to support students to develop from being "cue-deaf" to "cue-seeking", acknowledging all the current influences upon the field that a student might have to take into account.

5. ASSERTING TACIT KNOWLEDGE

In his analysis of the language of journal articles in biology, sociology and literary criticism, Bazerman (1981: 363) argues that readers expect certain knowledge and attitudes to be inherent in texts, and that the texts of particular disciplines will convey systematic differences in this respect. Gerholm (1990) terms them the tacit "rules of the game". Becher (1987b: 48) identifies different values expressed in texts from different disciplinary backgrounds: there is a certain self-consciousness about professional procedures in theory-dependent fields such as sociology, while in history there is little concern with theory, though craft and technique are emphasised. The attitude is different again in physics, where there is an assumption of confidence about such issues as the status of knowledge and the validity of method. Gerholm (1990: 267) argues that there are tacit conventions about "what counts as a relevant contribution, what counts as answering a question, what counts as having a good argument for that answer or a good criticism of it."

It is possible to articulate and discuss many of the rules of the linguistic game of the doctoral thesis, but there are also rules that remain implicit, conceptual but not concrete, and not open to being verbalised. This is consistent with Reber's (1997) view of human cognition and the importance of implicit learning. It is certainly the case, though, that successful doctoral theses illustrate mastery of the rules of the linguistic game. How do doctoral students come to know them?

Gerholm (1990: 270) distinguishes between the importance of the tacit knowledge grown out of long experience in the discipline and tacit knowledge generated as the student tries to make sense of his or her graduate studies. How students learn to assert their tacit knowledge is still open to argument in the psychological literature, though it is possible to examine how students convey the fundamental, substantive knowledge of their fields and how they convey the new knowledge they wish to add to it. The following example, from a doctoral thesis in geology, illustrates the portrayal of substantive knowledge that is tacitly understood to be known collectively by scholars in the discipline:

The banding which characterises BIFs occurs over a wide range of scales from thicknesses of several metres down to fine bands discernible only with a microscope. Because the various scales of banding form a suitable basis for descriptions of compositional and textural characteristics of BIF sequences, a reference nomenclature based on easily visualised scales of banding is necessary. Trendall and Blockley (1970) proposed such a scheme based on an apparent hierarchy of macrobands, mesobands, and microbands (terms originally introduced by Trendall, 1965a) but further work (Ewers and Morris, 1981; and this study) has shown that these three terms as originally defined do not provide a sufficiently comprehensive set of names to describe all band types. The principal shortcomings of the macroband-mesoband-microband nomenclature of Trendall and Blockey (1970) are as follows: . ." (McConchie, 1984: 7)

The writer builds upon existing research to argue for a *reference nomenclature* that he then shows to be inadequate, thereby paving the way for asserting the new knowledge to come. In building the argument, the writer has confidently assumed that the audience will understand the implications of references to *fine bands discernible only with a microscope*, will acknowledge an *apparent hierarchy of microbands* and will accept that there are other *band types*. However, there is further explanation of the origins of the banding terms which the audience might not already know. Each of the knowledge claims is supported by the tacit assumption that the reader shares the same substantive knowledge of the field, and so the author is on safe ground in making the assertion about shortcomings towards the end.

Another aspect of the above text, which demonstrates the writer's mastery of inexplicit writing conventions, is that the reporting format dominates. There is a keen use of technical terminology and an emphasis on classificatory, taxonomic organisation of the material. Furthermore, the writer avoids making overt the principal shortcomings about existing classifications or findings. This permits the audience to make judgements about the argument according to the conventional expectations of science writing, where information is expected to be reported without bias, even disinterestedly (see, for example, Bazerman, 1981; Myers, 1990). The overall effect is to emphasise concreteness and the writer's objectivity in asserting the argument.

5.1. Broader Conventions

The above example, though, does not illustrate some of the current influences upon field-specific language that may have to be taken into account by the student. The growing list of stakeholders in doctoral education, such as employers, industry and commerce, have their own expectations of good writing that are increasingly being extended to doctoral theses as examiners are increasingly drawn from outside the academy.

In demonstrating field-specific, tacit knowledge, thesis writers judiciously use terminology and citation to assist in making reference without qualification or explanation to the terms that have become 'universal' in the field, while choosing to describe newer or less well accepted terms, concepts, processes or approaches. This technique, especially important in humanities and some social science fields where knowledge is interpretive, establishes a common language with the audience and serves to build a framework from within which new knowledge can be asserted. In fields where knowledge is less concrete, the creation of a common language with the audience is essential. The associated self-consciousness is also an inherent part of the expression of disciplinary knowledge, as this further text from the Proudford thesis on education (quoted above) shows:

Whereas the literature conceives of the leader as one who rallies people to "a common cause" (Sergiovanni 1984, 8), and "helps to articulate, define and strengthen . . . endearing values, beliefs and cultural characteristics: (Duigan and MacPherson 1987, 55), the principal at North, as noted earlier, accommodated to the dominant values of the local community and set about building a school culture which reflected these values. While, at one level the principal can be regarded as a culture builder, the concept of leader as culture builder needs to be set within the context of the argument developed by Bates (1987). . . . (Proudford, 1992: 202)

Here the writer self-consciously asserts a personal view but does so in a way that is justified by the already established framework of thought. The doctoral student is demonstrating an understanding of the metaphors concerning *school culture* and the concepts of *educational leadership*. The assertions of previous researchers are directly used as part of the argument to persuade the audience of the interpretation being advanced, demonstrating that the writer understands the values of the target audience.

In the humanities, the expression of field-specific, tacit knowledge consists largely of deploying acceptable craft and technique. To the outsider, the following text, from Australian literature, confidently exemplifies the craft of the field and invokes images and concepts that seem part of its consciousness and knowledge base:

This pre-amble does have a design. Even though I am trying to trip up the journeyer, the navigator, the surveyor, I am not just idly playing games. I am trying not to contradict myself, not to present myself as the questing hero, the explorer, the discoverer, about to launch into a pursuit of the truth. These are figures who feature a lot in the following pages. We are working with metaphors here. I am trying to heed the warning of Edward Said: 'Our worldly-wisdom has been applied, in a sense, to everything except ourselves.' He is referring to the trap of the conquest whereby the old journeyers, deconstructed, are replaced by the new, who may be just as capable as the old of adopting the colonialist position, perhaps to a different territory. This thesis is informed by a cautious attitude towards both the traditional realist traveller/explorer and the more self-conscious non-realist manifestations of the figure; one if its interests is to examine the ways in which the motif is employed to both support and subvert reflective theories of language. (Gillett, 1990: 5)

This writer adopts strong metaphorical language, utilising elements of widely accepted terms of the field (*deconstructed, colonialist position*), thereby demonstrating a relevant tacit knowledge lexicon. A strong narrative asserts the writer's perspective, which is expressed as new knowledge. Here the doctoral student shows a tacit understanding of how to convey the existing knowledge of the field as well the new knowledge acquired as a result of her doctoral studies.

These examples of doctoral thesis texts are not special in their capacity to convey sophisticated epistemological understanding and social conventions. What is remarkable about them is that they provide evidence of the tacit learning and understanding first articulated by Polanyi (1961). In the years since *The Tacit Dimension* was published, the psychological literature has yielded a plethora of studies off human cognitive processing, distinguishing between the reasoned, conscious form and the subconscious, affective form that may be impossible to verbalise (see, for example, Epstein and Pacini (1999). While many researchers have investigated the cognitive processes involved in tacit learning, none to date have appreciated its sophistication, even though it may be argued that its evolutionary primacy makes it autonomic. Epstein and Pacini (1999: 477) explain: "The rational system develops in relation to the acquisition of language. Language permits a higher lever of abstract reasoning . . ." Doctoral students do know what it is that they have learned, because what they have learned is performative: they can perform discipline-specific writing at the highest level of sophistication, but they cannot describe the rules of the game.

6. EVIDENCE OF SOCIO-SEMIOTIC LEARNING

We have examined in some detail examples of successful doctoral texts and discussed how students have approached the task of writing up their doctorates. How do doctoral students come to know these different conventions when they are not explicitly taught? The answer, which lies somewhere in implicit cognitive information processing, is a key element in the socialisation of the student into a tribal way of reading, making and reporting knowledge. Knowing this, we can appreciate the need to provide opportunities for students to imitate, practise and obtain feedback on their writing. But while supervisors variously feel a responsibility to provide feedback on drafts their efforts may not be strategic in terms of trying to identify patterns, conventions and formats of expression that 'work' and to highlight those which do not.

The very notion of codified language being heavily value-laden provides a basis for examining further the ways in which doctoral students might be assisted to develop adequate writing skills. Since learning is a social phenomenon, it is worthwhile looking for opportunities in which students might receive feedback on their drafts, as well as those where scholarly writing may be compared and contrasted.

In this chapter, the nature of discipline-specific language has been reviewed in terms of its importance in representing a range of less than obvious disciplinary norms and values, the conventions for which have to be tacitly learned and mastered by doctoral students. But a note of caution should also be recorded. In his analysis of disciplinary language, Bazerman (1981: 579) concluded that:

We cannot even begin to speculate on what uniformities with what variations exist within disciplines or whether patterns of differences emerge among disciplines until many more examples have been examined and statistical indicators found to test the generality of conclusions.

His advice may have been prescriptive, but it points to the complexities of language registers which are constantly changing. Today the student not only has to understand the uniformities and variations in their own discipline, but also those of any related fields of knowledge upon which they draw, including communities of practice such as the professions. This chapter has been concerned to illuminate the sophistication of both the subconscious and the conscious learning that doctoral thesis writers demonstrate, and in doing so, to provide some insight into how supervisors might increase the range of opportunities for their students' learning. The discussion also throws light on the significance of Gerholm's (1990: 266) identification of "the ability to define the situation correctly and use the type of discourse required by that very situation", beside which, he asserts, other forms of tacit learning "shrink into insignificance".

PART 3

FOUNDATIONS AND NEW HORIZONS

CHAPTER 8

ACHIEVING SOCIALISATION

1. LEARNING IN A CHANGING ENVIRONMENT

This chapter sets out to bring together earlier findings about opportunities for learning disciplinary values and conventions in the social setting for research to provide insights about the kinds of research settings that are more likely to lead to timely, successful outcomes for doctoral candidates. We have seen how induction to disciplinary cultures takes place by immersion in those cultures: in these settings, students develop intellectual and perhaps also social rapport with established scholars and other doctoral students. They compare and contrast their intellectual endeavours with these people and begin to develop an intellectual identity, together with a set of commitments to certain values, methods, traditions and ways of thinking. And yet, these knowledge-making cultures are constantly evolving, developing and changing to the point where, as Becher and Parry (2005) point out, Mode 1 and Mode 2 knowledge co-exist in the settings where knowledge is made, and therefore, in the settings where doctoral research is carried out. There is, therefore, a growing dimension of uncertainty in knowledge-making settings, particularly as knowledge-making activity rapidly expands across the developed world.

Doctoral programs provide a wide range of socialising opportunities through which the traditions, conventions and values of a discipline may be learned. In newer, Mode 2 knowledge areas, including transdisciplinary and applied areas, having less well established traditions and drawing at will from divergent knowledge fields and methodologies, would seem to be less capable of providing well-planned socialising opportunities for novice researchers. Many field-specific characteristics take time to be noticed or documented, and traditions relating to scholarly communication and networking as well as the documented influence of exemplars in the field, take tome to be valued. Such characteristics and traditions will therefore not be immediately obvious to established scholars, much less to their doctoral candidates.

The trend towards Mode 2 knowledge indicates that doctoral study programs are becoming increasingly individualised. In this context, appropriate socialising opportunities in the broader scholarly community are even more urgent as candidates negotiate schools of thought, methodological options, the influence of like-minded exemplars and the grafting on to a core discipline many other knowledge strains. At the same time, it needs to be remembered that, though highly mutable, the character and epistemological values of specific fields inherently constrain the ways in which knowledge production and reporting can occur: codified language still needs to be accepted and understood by the examiners – the intended audience.

We have seen in Chapters 5 and 6 that it is often through the student's initiative that socialising opportunities in the cosmopolitan arena are sought. The demonstrably *ad hoc* approach to providing such opportunities by institutions, particularly in certain knowledge fields, is no longer sufficient or appropriate, given the greater uncertainty for students, particularly in novel or transdisciplinary fields. There is, of course, the danger that non-science fields of knowledge will shrink because they take longer to accomplish and are more expensive to resource. The evidence also shows clearly that the learning process during doctoral study is largely subconscious and yet it is highly formative, so opportunities for comparing and contrasting approaches, values, traditions and methods need more than ever to be strategically made available. Given the greater financial burden on doctoral students in most countries, this imperative should become more and more obvious. Yet there is to date no evidence of a strong movement within the humanities and social sciences to address this issue.

The capacity to demonstrate mastery of the core elements of social and epistemological properties of disciplinary cultures is fundamental to attainment of the research doctorate. Once this acceptance is reached with the audience, there is the capacity for the thesis to influence accepted epistemological characteristics, thus opening the knowledge base to change. Here we have a useful illustration of the metaphor of the "scientist having to rebuild the ship while on high seas" (Cartwright et al., 1996) which Nowotny, Scott and Gibbons (2001:178) take to encompass both cognitive content as well as the more heterogenous, ambiguous constitution of reliable knowledge required to withstand the tempestuous seas of uncertainty in contemporary society.

1.1. Disciplines and Change

In much of the recent research on doctoral study, disciplinary characteristics, traditions and expectations are widely acknowledged, but they are treated as though they are generally explicit elements of the doctoral research setting. But there is a need for the complexities of new knowledge making to be acknowledged if research training is to be effective and efficient. By Delanty's (2001:152) account, " . . . the conditions for knowledge production are no longer controlled by the mode of knowledge itself . . . (generating) . . . new knowledge fields" that are more applied and Mode 2 (Gibbons et al., 1994) in their production.

The more focused the social interactions are towards learning disciplinary and field-specific lore, the better the induction of the student. In these circumstances, the student is more likely to have greater access to and possibly commitment from, scholars who are already part of the particular knowledge landscape; thus too, they are likely to have available to them better benchmarking opportunities to compare and evaluate their own performance on a range of skills, capacities and attitudes.

In the increasingly applied contemporary context of research, it is curious that policy makers and institutions alike largely ignore these features of knowledge making and reporting. Instead, students are typically channelled into collective research enclaves that are artificial social communities constructed by institutions to address intellectual and/or epistemological isolation that cannot be resolved by an uninformed change of the social context or networks of doctoral candidates. Most frequently, institutions seem to be keen to collectivise the research enterprise, even where this is not appropriate. The artificial collectivisation or 'sciencification' of the research enterprise is at the very least confusing, adding to the potential for students to be distracted from their core disciplinary focus and to be confused about the conceptual connections they might graft to their particular disciplinary core.

Given the multiple publics and multiple sources upon which new knowledge draws, it is not surprising that Nowotny, Scott and Gibbons (2001) have questioned whether the disciplinary 'core' is empty, alluding to the fundamental change in the nature of knowledge in the twenty first century. But such a notion is far too simplistic, for it ignores all the evidence pointing to a socio-semiotic conception of knowledge, both in terms of its making and its reporting, which, at the level of the doctorate at least, are not separable conceptions. As the research enterprise becomes more specialized and research enclaves less densely populated, the need for knowledge makers – including doctoral students – to identify themselves in terms of core values, methods and traditions is all the greater. While national systems continue to treat research and the doctoral research enterprise as an undifferentiated learning experience with a uniform set of outputs, the disciplinary core giving focus to knowledge-making will continue to be overlooked, masked by the stunning rapidity of change in knowledge production in contemporary society.

1.2. Elements of Induction

The socialisation of doctoral students to disciplinary communities has been described as involving the development of a sense of identity with and personal commitment to the disciplinary field (see, for example, Becker and Carper, 1956; Geertz, 1983; Henkel, 2000; Delamont, Atkinson and Parry, 2000), especially for those seeking a career in higher education. Two essential elements in the development of a disciplinary or academic identity are the social setting for doctoral work, including the academic department or research group or graduate school, and the broader international research community, including its traditions, conventions and values. These two elements, echoing the concepts of 'habitus' and 'cultural capital' (from Bourdieu, 1977), constitute the social environment for doctoral study and a rich framework for learning epistemology and disciplinary values. But the richness of the framework depends upon the strength of its disciplinary foundations and the influence and status of its social actors in the broader knowledge community.

1.3. Immersion in the Research Culture

Immersion in a disciplinary setting during candidature provides fertile ground for shaping disciplinary identities and values. Explicit departmental and supervisory practices set identifiable parameters for students about how research is conducted and reported and about how the disciplinary society works. These parameters are important. Through them, an understanding of the influences upon the existing knowledge base can be refined, and the conventions for undertaking research and for expressing new knowledge are learned and practised.

In the experimental sciences, for example, students are initiated into a culture where teamwork and collaboration are valued. The rewards for investment in this culture are shared achievements – which are highly prized. Paradoxically, these values usually lead to a strong commitment by the student to a particular research question, even though frequently the student may have had little or no say in choosing it. The social setting of team-based science demands a range of personal commitments that engender a sense of responsibility and commitment to research outcomes. Newly recruited doctoral students readily adopt these values. The neatness of fit of a new recruit works two-ways because the willing and enthusiastic student is rewarded and encouraged in this setting.

Immersion in the different settings of the humanities where research specialisms are more individualistic sees doctoral students working autonomously from the beginning, sometimes with a disconcerting sense of isolation and of having to confront intellectual uncertainty. Making intellectual gains in these circumstances is hard earned though it inculcates a strong sense of individual achievement and ownership of the research topic. In these settings, students value their own intellectual achievements highly, and this high value is consistent with the increased levels of intrinsic motivation long reported among humanities students, compared with the vocational aspirations more frequently reported among science students (see, for example, Powles, 1984). Making intellectual gains is probably harder work, a view put forward by many non-science supervisors in the investigation. The relatively fewer resources and opportunities to network, the more sparsely populated and scattered demographics and the uncertainties in the knowledge base of humanities specialisms, mean that there are fewer opportunities for socialisation to the disciplinary culture. Humanities supervisors typically report that departmental seminar programs for doctoral students are vital. Students, of course, want affirmation, stimulation to broaden their intellectual horizons, and opportunities to practise the rules of the disciplinary game. Of course, they cannot achieve these objectives without feedback from scholarly authorities, and so the involvement of senior scholars is important, as is the quality of the critique provided. Not surprisingly, in settings where senior scholars were largely uninvolved in seminar programs, students likewise were ambivalent about them.

The research cultures in which social science students are immersed express values that are different again, though there is so much diversity across individual fields that it is impossible to make generalisations about processes of socialisation or induction. It is possible however, to highlight some of the related issues and concerns of students so that opportunities for effective socialisation might be identified and supported.

Characteristic of most social science specialisms is the high value placed upon analytical expertise and well-developed theoretical stances, features that provide an explicit set of parameters for new recruits in making knowledge gains. For doctoral students the messages are unequivocal: demonstration of appropriate skills and techniques is imperative, and scholarly exchange with those whose theoretical or methodological approaches are similar is vital. Doctoral students need opportunities to develop and enhance their skills and expertise as well as opportunities to communicate with notable scholars with whose research they identify. Most important, however, are the benefits derived from networking with scholars in the broader arena: it is from socialising with these scholars that disciplinary identity is developed and focus is determined.

However, it is important to acknowledge the individualistic nature of social science research, which typically does not lend itself to collective inquiry. The range of theoretical perspectives and methods of inquiry cover a very wide range, so students who choose their own topics may well have little in common with others in the same organisation. Where they do, the synergy is likely to be very broad, such as in theory building which includes a particularly influential scholar, or using a post-positivist methodology such as grounded theory. In these circumstances it may well be distracting for students to be required to undertake prerequisite coursework, or to co-publish during candidature, unless of course the diversion from doctoral study marks out territory and is relevant to the research at hand.

Immersion in typical research settings provides an organising framework of disciplinary values, traditions and expectations in which doctoral students begin candidature and start the process of making, and reporting, knowledge. In these settings, the cues given to them are clear and unambiguous. Supervisors and significant others in the research setting – including the broader, cosmopolitan research culture – play key roles in providing opportunities for students to imbibe cues such as these, so it is essential that students learn to navigate their way around the different intellectual territories with a clear focus and purpose.

2. SUPERVISORY RELATIONSHIPS

There are other administrative rationales for organising doctoral study programmes. Vilnikas (2000) used a management model to derive eight operational types, and nine different approaches to doctoral supervision. Within this view, a supervisor might be considered to be the innovator, the broker, the producer, the director, the co-ordinator, the monitor, the facilitator, the mentor and the integrator. All these types, however, ignore the nature of the intellectual relationship between the supervisor and the doctoral candidate, and how this is constrained by cognitive and social features of the knowledge base. Where doctoral students form a fundamental commitment to the rules of knowledge making and knowledge reproduction, there is a shared intellectual commitment to the completion and reporting of the research – and this is a widely recognized ingredient for successful candidature. There are qualitatively different kinds of relationships in different disciplinary settings between supervisors and their doctoral candidates, and also among colleagues and peers, which reinforce distinctive disciplinary traditions and contribute to effective induction.

2.1. Intellectual Rapport in Science

In science areas, doctoral students are treated very much as raw recruits, given step-by-step direction at first and frequently left to resolve low-level questions about their research by seeking help from technical and support staff, which shows them their place in the hierarchy of scientific research culture. Though there are frequent meetings between students and their supervisors, especially in team-based settings, and though some collaborative teams may engender rich sources of friendship and personal commitment, there is a tradition of distance between supervisors and their students that reflects the nature of pecking orders in science disciplines. The priorities in science first and foremost concern the rapid production of knowledge outputs and the characteristic kinds of working relationships formed during doctoral study reflect these priorities. But the emphasis upon teamwork and shared outputs also engenders a set of shared commitments that benefits the student as well as the research team.

2.2. Intellectual Rapport in Interpretive Fields

The emphasis is markedly different in humanities specialisms, where doctoral students are treated more like junior colleagues whose intellectual outputs are respected and valued. This respect makes for a more collegial approach than in other disciplines, where it typically forms a platform for commitment to the research even where research interests are not aligned. A good rapport between supervisor and student is essential because it provides the primary source of affirmation about particularistic research where intellectual isolation is the norm. It is noteworthy that humanities students who did not form a close rapport their supervisors invariably spoke about a scholarly 'guru' whose mantle they worked at adopting. This kind of close intellectual relationship with an influential scholar also provides a primary conduit to networks of scholars in related fields, who in turn provide parameters for student to further situate their research in relation to the broader discipline.

In the eclectic fields of social science research, where there is relatively little collaborative research, students are more likely to be treated respectfully, but with a degree of professional and status-oriented distance, from supervisors. In these settings, communities of scholars tend to be less hierarchical than in the sciences, but there is a risk of differences in values giving rise to a lack of trust and commitment towards by the supervisor. Internecine disputes and secular interests lead to strong alliances among allied scholars, and it is within this social setting that students must expressly identify with a particular school of thought and intellectual sub-culture. Supervisors (or a significant scholar who serves as the equivalent) are instrumental in guiding the process of identification with like-minded scholars and schools of thought. As we have seen Chapters 6 and 7, the result of these intellectual conditions is that, unless there are theoretical or methodological disputes, a comfortable intellectual rapport usually develops.

2.3. Rapport and 'Professional Relationships'

While it is widely recognised that the development of a good rapport between supervisor and student is an important ingredient in success for most students, this topic has not been given enough empirical attention. When asking supervisors about their relationships with their students, there was uniformly a desire to express the 'professional' nature of the relationships formed. Students, though, were less conscious of crossing professional boundaries and were more open in their descriptions of personal dynamics with supervisors. What was clear in all cases, though, is that a strong intellectual rapport, which is necessary for joint commitment to the successful completion of the research project, is based on shared values and a shared set of commitments in the field.

Taking a decidedly negative view of the supervisor-student relationship, McWilliam, Taylor and Singh (2002) adopted a management model of supervision, outlining the importance of managing the risk of a poor supervisory relationship. These authors exemplify risk as the "lecherous professor" and the "harrassable student". Other authors describe the changing nature of relationships in recent times (Linden 1999, for example). For some supervisors, rapport is essential. Some boast that they can supervise in any field, because the vital ingredient in the process is a commitment to successful completion. Others report that a commitment to the completion of the research is the key. Yet others still report that intellectual rapport is only possible when supervisor and student share an interest in a particular topic. For students, however, their need for intellectual rapport is underscored by their need for validation, for which trust is an essential ingredient. At the heart of a good intellectual rapport between supervisor and student is a shared identification with the core values and aspirations of the discipline or the specific field, recognising that in individualistic fields of knowledge, some dissonance is inevitable.

To complicate matters, intellectual rapport is likely to be more problematic when more than one supervisor is involved. There is the benefit to the student in being able to draw upon a broader range of expertise and in having critical appraisal from more than one source – and possibly more than one specialism. However, in more individualistic fields, supervisors rarely share methodological approaches or theoretical frameworks. More complicated still is that co-supervisors may be drawn from industry, commerce or the professions where communities of practice may be dissociated from theoreticians and the research setting within universities. All of these differences in values and perspectives have to be contextualised and understood by the doctoral student.

3. THE CONTINUING PROBLEM OF COMPLETION RATES AND TIMES

An ongoing concern in many national systems of higher education is that completion times and dropout rates in humanities and social science-based fields have not improved over several decades, despite governments intervening to redress this situation. Apart from the cost of lengthy or unproductive candidature, there are questions of accountability associated with this problem. A widely adopted intervention has been to reorganise doctoral research and study programs within institutions into enclaves where the collective research culture is intended to compensate for intellectual, social and emotional isolation.

However, Neumann (2002) points out that in the context of the Australian system, three interventions that have been introduced to improve doctoral study programmes have not made a contribution to their government's stated "knowledge nation" objective. Those interventions were to redress completion rates and times by: limiting cost by restricting stipends to the minimum period of candidature regardless of discipline; by concentrating the research enterprise into key areas, and by providing 'relevant' coursework to doctoral students to make them more employable.

But these interventions ignore the epistemological conditions constraining the making and reporting of new knowledge; they do not recognize that some research quite naturally takes longer to complete to a high standard; they ignore the epistemological conditions constraining individualistic and collaborative research; and they make the rather great conceptual leap that a philosophical doctorate is a licence for work instead of a licence to do research. Of course there is as yet no evidence that the interventions have resulted in cost-savings for government.

In order to understand why these interventions have so little impact, we must look at the nature of the disciplines concerned. Knowledge manufacture in individualistic fields is a "top down" process where the audience has to be convinced of its worth and persuaded about its insights and their value. Not only is it highly likely to be individualistic, but students are also more likely to be older, juggling work and family responsibilities, female and perhaps also already have an established professional career. Typically, science students in their early twenties who come into doctoral programs directly from honours degrees have many advantages, including the collective nature of research, densely populated research communities – and often – a stipend or scholarship. Identity formation in the latter case is bound to be fairly straightforward. Those frequently older, mature-age humanities students and career-minded social science and applied professions students are immediately set at a disadvantage, with their work, study and family responsibilities. With governments in most developed countries espousing the mantra of lifelong learning, policy analysts might well question the ways in which research in individualistic and transdisciplinary fields of study might be better supported.

3.1. Developing Disciplinary Savvy

There has been a tendency in recent times for researchers and policy makers to refer to research education as training. That development begs the question of whether the attainment of a PhD be reduced to skills training. Whether it is possible to identify core skills training for undertaking a PhD is arguable, but the trend towards attempting to identify them does diminish the achievement somewhat – particularly when there has been a long period of sustained scholarship underpinning its achievement. The development of a disciplinary identity is required, an achievement which is considerable. Some students are conscious of learning discrete norms in this way while others seem unaware of doing so. Students use terms such as 'collecting trophies', 'climbing the greasy pole' and knowing the 'Mafia' to refer to the means by which they learned inexplicit or tacit norms. While this is consistent with the notion of cue seeking described by Miller and Parlett (1976:144), these descriptions suggest that many students consider that their having discovered the tacit 'rules of the game' is almost dishonourable. It is not coincidental that a mature-age female history student commented: "When I started doing this, I thought it was really about writing a thesis. It isn't about writing a thesis at all. It's about politics, about being accepted, about being supported. I shouldn't say that, should I?"

Not all students feel this way. A biochemistry student, having just submitted a highly praised thesis, reported: "How did I do it? I just do my work and keep my head down and down and do what I'm told. And I learn from my reading. And of course, I know what to do when I get a crystal, because everybody knows that when you get a crystal, you write it up." Of course, this student is really making the same claim to working out the socially-related power structures in the discipline and working both cognitively and socially within their parameters, a point also illustrated by Kogan (2001).

Curiously, many doctoral students report believing for considerable periods during their candidature that they were not intellectually capable of completing a doctorate, though this typically comes from students in social science or humanities fields. A sociology student who was also lecturing in her department commented, "I thought I was just acting, pretending I could think at their level. I didn't feel I could actually do [the thesis]. I felt like I was tricking [my supervisors], especially for the first eighteen months. But later on, you get to know what's going on, and you begin to think, 'well, maybe I can do this. When I submitted [my thesis], I didn't even care what (the examiners) thought of it. I'd done it and I felt as smart as they were . . . I felt pleased with myself'."

Descriptions of students developing a tacit understanding of their learning context, in which they tend to emphasise the social rather than the cognitive elements of the learning situation, are highly consistent with empirical studies of unconscious learning. Epstein and Pacini (1999: 468) for example, found that "the experiential system responds to the overall context of situations rather than to isolated, abstracted elements". Moreover, they distinguish between automatic, effortless cognitive processing and intentional, effortful cognitive processing, in which some doctoral students were able to describe their cue-seeking behaviours. These distinctions are useful because they permit us to understand better how students learn from immersion in their particular research environments. So, while students may find the immersion overwhelming at first, they seem to adapt to fit in to the milieu and work within its values and conventions.

3.2. Induction to Complex Knowledge Domains

Students, particularly in non-science fields that are individualistic or particularistic, face increasing uncertainty in the context of applied and transdisciplinary research undertakings, as many observers have noted (Nowotny, Scott and Gibbons, 2001; Weingart, 2001). Certainly the unity of science is, as Knorr-Cetina 1999 points out, at the very least, questionable. We have discussed this development in several contexts, and yet we have seen that induction to a disciplinary "core" is pivotal to the focus and containment of doctoral research. And this condition for success is undiminished as the nature of knowledge transcends Mode 2 conceptions and its more public role as described by Delanty (2001: 74). Doctoral students have to develop enough disciplinary savvy to establish what constitutes valid or reliable knowledge in their field of enquiry.

Kogan (2001), commenting on changing knowledge and power patterns in society, drew attention to Polany's (1962) view that the validity of knowledge is enforced not by proof (and therefore the dictates of disciplinary authority) but by responsible judgement, and also to Popper's view that the power of science lies in putting its propositions to the test, rather than in endowment by authority. In Kogan's (2001: 20) view, community and lay perceptions of what applies and what works in terms of acceptable knowledge apply across the boundaries of the spectrum of knowledge. However, he also acknowledges that "official knowledge" at any time is powerful. And so it is with communities of practice such as knowledge communities, with their multiple reference groups and extended boundaries.

The importance of socialising with other members of a dominant or intellectually powerful professional or disciplinary group has been noted in existing research (see Bucher and Strauss, 1960; Becker and Carper, 1956; Kolb, 1984; Gerholm, 1990; and more recently, Delamont, Atkinson and Parry (2000), but the role of tacit or un-conscious learning in these kinds of communities of practice has not generally been assigned the importance it deserves. Since disciplinary communities only allocate the rewards of acceptance and status to those who live up to their norms, as Merton (1957: 642) has explained, induction to specific knowledge fields must largely be completed during doctoral study. At this level, the expression of new and existing knowledge is shaped at a sufficiently advanced level to meet the established expectations of the disciplinary community.

Students may not be consciously aware that by mixing with and learning from others in the research environment, they are in effect embarking on a process of induction into specialised knowledge cultures. But experienced scholars are tacitly aware of the importance of powerful intellectual influences. Indeed, academic departments have long taken account of these in providing resources to doctoral study programs, as we saw in Chapters 5 and 6. How does this translate into learning? Epstein and Pacini (1999: 463) argue that tacit learning " . . . is a relatively crude, albeit efficient system for automatically, rapidly and effortlessly processing information while placing minimal demands on cognitive resources. At its higher reaches, and particularly in interaction with the rational system, the experiential system can be a source of intuitive wisdom and creativity."

Merton (1973: 268–9) observed that disciplinary norms within the "normative structure of science" are "imperatives which are transmitted by precept and example, reinforced by sanctions and internalised." Disciplinary norms produce, or reproduce, a specific epistemological consciousness that is open to change as the knowledge base evolves. Supervisors, colleagues, peers and those in the broader disciplinary arena may facilitate this kind of tacit learning. The significance of field-specific socialisation which provides ample opportunities for tacit learning during doctoral study is widely underestimated, as is the role of significant others, those powerful intellectual influences who serve as exemplars to their novices, in a doctoral student's epistemological milieu.

4. TACIT KNOWING

Gerholm (1990) identified different kinds of tacit learning. For example, an intuitive knowledge of the essence of one's own discipline and of its relationship to neighbouring disciplines cannot easily be differentiated from the ability to recognise what Gerholm (1990: 267) terms 'valid arguments', 'telling objections', 'insightful questions'. Important among forms of tacit learning is the capacity to recognise conflicting norms and to know which norms are appropriate to certain circumstances and when to invoke them. These various forms of tacit learning constitute the *savoir faire* or political consciousness so widely recognised by supervisors and students alike as being essential to operating successfully in a disciplinary arena (see, for example, Becher, 1989a; Gerholm, 1990; Merton, 1973; Mitroff, 1974). Successful students learn what will be acceptable in a particular intellectual community and, within this setting, in particular situations. They are thus able to assert themselves confidently. Disciplinary know-how is a highly sophisticated skill that we can observe in practice by examining the typical research settings of particular fields.

A prevailing requirement of students in experimental and team-based science fields is to mark out intellectual territory as their own. While students learn that this imperative must be met, they also discover that there is much to be learned from others in the laboratory or team and much to be gained from patronage by the supervisor, especially when the supervisor is influential internationally. In several fields, supervisors reported contracting with their students to be included on at least one refereed publication as part of the 'candidature deal'. At the same time, while recognising some unpalatable aspects of such a contract, most students report that, on balance, they gain more than they lose from the practice. Generally these students argue that collaboration, and publishing with an established scholar, are ultimately of reputational value and not to be missed.

In team-based science settings, students are given the explicit message from the beginning that collaboration prevails; yet successful careers depend upon individuals marking out their own intellectual territory. Whether this widespread practice is justifiable or not, the dilemma it gives rise to for the student is consistent with the notion of counter-norms, as identified by Mitroff (1974). Students are confronted early in candidature with conflicting norms like this, and are able to learn from socialising with their peers in the laboratory environment how to deal with them. Those who eventually acquire the disciplinary savvy required to operate as independent scholars are able to judge appropriate times to assert their own ideas, either in print or by surpassing the knowledge of the supervisor in the thesis. To operate successfully in the research arena, both a foundation of disciplinary knowledge is necessary, together with an understanding of how to exploit new knowledge, in this case, when to assert independence, and when to acknowledge the contribution of others.

Students in the humanities begin candidature expecting to work in intellectually isolated circumstances. Their investment in the disciplinary area is substantial because they have to apply themselves to working autonomously before they have had the opportunity to confirm many of the conventions of the field through socialisation processes. While supervisors widely report providing strong support

to their students during the early stage of candidature, it is the students' independent intellectual achievements that provide their greatest sense of achievement
and disciplinary pride, as history students described so clearly. Students may not
even know they are initially in a 'sink or swim' situation while they negotiate a
delicate balance between demonstrating independent ability and seeking disciplinary direction from the supervisor. Yet, achieving an appropriately constructed
argument reflects a fine balance between, on the one hand, their independence and
individuality, and, on the one hand, their conformity to disciplinary conventions.

Counter-norms also prevail in the diverse range of social science specialisms,
where skill development and sound theoretical frameworks are reflected in supervisory
practices. At the same time, a feature of social science is the extent to which specialisms
are dynamic: constantly changing according to the influence of particular scholars and
paradigms, with some fields exhibiting intellectual fashions and the development of
cliques among scholars. Conflict gives rise to counter-norms that must be recognised
if students are to make choices in order to align themselves with appropriate schools of
thought and relevant scholars. One supervisor used the term, 'politicality', to describe a
quality students must develop. The tacit learning achieved by students in these settings
is both complex and considerable. On the one hand, they must assert the foundations of
their primary discipline in their doctoral research, as Parry, Delamont and Atkinson
(1994) have shown. On the other hand, they must discern and adopt theoretical
and methodological perspectives that are appropriate to the intellectual climate of
the time and field, and which perhaps also extend existing boundaries. This capacity
fits comfortably with Reber's (1997) description of the source of "intuitive wisdom
and creativity."

Juggling counter-norms is especially difficult to do in cases where students have
multidisciplinary or applied research topics because their focus may cross over
theoretical, methodological and ideological boundaries. In these settings, there is a
greater need for intellectual affirmation, though it may well be more difficult for
supervisors and significant others to provide it because, as we have seen, paradigms
can be individualistic. It is not surprising that, in these fields of research, doctoral
students are more likely to take longer to complete their study programme.

5. REPRODUCING SPECIALISED KNOWLEDGE

5.1. Informal Communication

Important opportunities for tacit learning occur through formal and informal communication between doctoral students and their colleagues and peers. Through
the various channels of scholarly communication, including publishing, attending
conferences, contributing to departmental seminars and engaging in informal discussion in the department, students learn a range of explicit and inexplicit values and
conventions about how knowledge is reproduced and exchanged in particular
milieux. In science, for example, publishing articles draws attention to the accretive
nature of knowledge and to the importance of marking out intellectual territory.

It also draws attention to specific reporting and citation formats, modes of argument, and avenues for asserting intellectual bite and for signalling collegial support for ideas advanced.

Publishing during candidature is not nearly so strong an imperative in many fields within the social sciences as it is in more competitive disciplines including the experimental sciences, and so informal channels of scholarly exchange assume greater importance because they provide alternative opportunities for learning and practising knowledge reproduction. Supervisors in social science fields, and particularly in economics and business fields, report helping doctoral students to get to know people within the department and in the broader scholarly networks. They argue that through opportunities such as these, students can learn about the influence of certain scholars, about the importance of keeping abreast of very new developments in their fields and about the boundaries of acceptability. Provision of these opportunities, though they may be limited, is seen as particularly important in newly emerging and transdisciplinary fields where epistemology may be fragile and embryonic.

Tacit learning by informal communicating with others is conditioned by a different set of circumstances in the humanities, where, owing to the highly personalised nature of research questions, individualistic values prevail. Here students have to establish that their argument is worthy of engagement, and this requires a certain level of competence in the relevant theoretical perspectives and developments. In these settings, it is not surprising that supervisors expect new recruits to demonstrate such competence in reproducing ideas from the beginning. Once a level of intellectual respectability is established between the student and supervisors, colleagues or peers, these individuals come to feature strongly in providing affirmation and critique. But these are not simple relationships.

What to share and how to do it are critically important conventions for students to learn in disciplinary cultures where it is antithetical for supervisors to 'interfere' with students' writing. A fine balance between seeking affirmation of knowledge claims and of asserting independence must be struck in developing a personal perspective. Success consists in being able to learn from others how this may be done and what are the discrete boundaries of acceptability that should apply.

5.2. Formal Communication

Not only is informal communication with others essential for the tacit learning of cultural mores, but there are also differing rules about the formal reproduction and exchange of knowledge. Students who publish collaboratively with supervisors and experienced scholars in team-based scientific fields must learn the usual codes of practice in the authorship of publications. Supervisors and students alike agree that these codes exist and that they vary from field to field, though the organisational setting of the department may also be an influence on typical practices. At one extreme, it is the norm for the student to author solely any publications completed during candidature, while at the other extreme, supervisors insist on being named as co-author in all publications completed during candidature, at times regardless of their input. In most situations, students learn to accept a junior role in the publication of knowledge until they have built up enough credits in the field to take the lead or to operate independently. Their credits in the field may reflect mastery of the conventions of the discipline, but how they are acquired demands a good deal of disciplinary savvy, in particular, knowing one's place in the fabric of the scholarly networks.

Students have to learn similar lessons in fields where publishing during candidature is not necessarily expected. In the humanities, for example, students learn to identify exemplars in their fields and seek them out for affirmation and intellectual support. Many humanities students report learning to identify the status distinctions and intellectual influences operating at conferences and meetings of scholarly networks, and using this information to better work their way into the intellectual landscape. Yet others describe how networking opportunities enable them identify field-specific alignments and intellectual rifts, so they can develop an understanding of who the key scholars are and the kind of influence these scholars have on theoretical developments.

In committing to doctoral studies and to learning in specific intellectual settings, students are making significant investments in their disciplines, for which they expect to obtain the rewards of affirmation, encouragement and, ultimately, acceptance. In the process, students make a commitment to producing and reproducing knowledge in certain ways. The sophistication and complexity of this task cannot be overstated.

6. DEVELOPING DISCIPLINARY IDENTITIES

While doctoral students must abide by disciplinary norms and values, this does not preclude them from expressing individuality or non-conformity to some degree. In fact, the advancement and development of disciplines depends upon scholars extending the boundaries of acceptability in accordance with the advancement of new knowledge claims. It does not, however, fall to doctoral students to champion this role, or to take large risks by departing too far from established norms. But the capacity for dynamism is important because it counters any claims we might make of academic disciplines being prescriptive.

Delamont, Atkinson and Parry (2000: 3) concur with Fulton (1996) that there are dimensions of contrast in the academic profession other than disciplinarity, such as rank in the profession, conditions of service and local academic and organisational cultures. While local and organisational cultures are certainly relevant, it is clear that identification with a specific discipline-based subculture is most important to successful doctoral study. In reality, the doctoral student presents a thesis to a small number of examiners, drawn from an international pool of experts in the field. These examiners are bound by their own disciplinary values, traditions and expectations in their assessment of the contribution of the thesis to its knowledge base. As a result, doctoral students who aspire to membership of specialized knowledge communities must learn to make new knowledge in ways that are acceptable and appropriate in those communities. They must also be able to report it conventionally, that is, using for the most part the accepted conventions in the field.

7. ORGANISATIONAL FRAMEWORKS FOR LEARNING

For students learning to 'get on' in their field of research and to make worthy contributions to the knowledge base, the complexities of the learning objects are immense. The kind of socially based activities available to doctoral students provides avenues for learning and for extending their knowledge in both conscious and unconscious domains. An advantage of socially-oriented learning is that it is diverse, constantly evolving and so it is well suited to triangulation and comparison. It also permits navigation through a community's dynamically evolving knowledge base. So, how do the social settings for doctoral research measure up to effectively providing sufficient opportunities for learning at the level of the PhD?

The overwhelming view of doctoral students is that their links with or networks into the appropriate international discipline-based field of researchers are more formative in achieving their doctoral objectives than their connections and status within their own departments. This view does undervalue the role of the academic department or research centre in supporting doctoral students or their research. The organization of the departmental setting provides the physical setting and resources for the research, and for close-at-hand feedback about progress and about thesis writing. However, doctoral students, by the time they come to 'know the ropes' towards the end of candidature, openly recognize that it is a select few scholars, drawn from the international scholarly community, who will assess the value and contribution of the thesis to the field.

In order to achieve a thesis of a standard that demonstrates clearly that they do 'know the ropes', doctoral students, within their departments or research centres, need to develop cultural savvy. They need to know how to access the financial, human and intellectual resources required to seek the intellectual supports they need for affirmation and to use their initiative to use colleagues in the department as a springboard to the broader scholarly networks. A geology student articulated the distinction between the organizational setting of the department and the broader, cosmopolitan scholarly networks: "We're in a good situation here (in this department) . . . when you get a (doctoral research) topic, you get instructions about the people in the department. We get to travel and go to conferences, but you have to work your way around the department to get these things. On the other hand, there's an entirely different game out there. There's an American group and a German group and then there's our group . . . Basically, you have to (be) credible in those circles, and I've done that with publishing." Focusing on socially based learning in these two settings, a clear distinction between what the organisational culture of the department can offer, and the need to be accepted by and learn from the international research community is being made. In doing so, this student demonstrates a tacit understanding of departmental etiquette and disciplinary lore.

But this student's experience exemplifies experimental science, where the scholarly networks are densely populated and where research funding supports discovery and the exchange of ideas. As a result, the student is enabled to interact with those networks on as many levels as he chooses: formally through seminars, conferences and colloquia; and informally through the attendant social opportunities, and by practising the reporting of new ideas and receiving formative feedback, guidance and affirmation. In addition, this student is part of a team, so the research effort is a collaborative one in which the quality of the student's doctoral work is crucial to other team players. By comparison, the lone scholar in an unfunded, interpretive area of research is likely to have fewer natural advantages in both their local and cosmopolitan research settings.

7.1. Disciplinary Etiquette

One of the ways in which policy-makers and institutions have set about overcoming the kind of comparative disadvantage of interpretive fields of study is to impose the team-based model of collective science upon students – even though it is wholly inappropriate for students in fields ranging from professional to interpretive. The kinds of skills and capacities needed to build confidence in the doctoral researcher in a highly particularistic field of knowledge are neither based upon shared perspectives on phenomena, nor are they accretive and subject to shared contribution. Indeed, students in particularistic fields of knowledge need to make a commitment to independence from the beginning, as well as to exercise the initiative required to seek out feedback from important intellectual influences in the field. The individualistic nature of the research enterprise governs the solitary nature of theory building as it does the development of individualized perspectives and reporting formats. The kinds of skills needed to succeed in the face of intellectual uncertainty are distinctly different from those in collective areas of experimental science, where there is always the safety net of team members' reputations, and therefore input, to draw upon.

Doctoral candidates report considerable difficulty and confusion in navigating their way into the appropriate social etiquette for their fields, especially taking into account the ever-increasing trend towards transdisciplinary research. And yet, for these candidates, it is all the more vital to their success that they do so because their reporting of new knowledge needs to meet the expectations of examiners.

From less explicit elements of the setting for research, lone students in individual-istic fields must learn established modes of acceptable behaviour such as those evident in the codes of conduct governing thesis and article writing. For example, we saw in Chapter 7 how authorship is the focus of citation in the humanities. Yet the etiquette of disapprobation, as Becher (1989a) and Bazerman (1989) have shown, is highly field-specific. Students need to navigate their way through different sets of conventions attached to different theoretical traditions. We have also seen that super-visors are generally unable to articulate conventions such as these, and yet the student still must learn to manipulate them. A student in philosophy, for example, may have to 'knock someone else's argument off' in order to assert a new, more insightful perspective. In a doctoral work, though, as students openly attest, a student cannot risk offending a potential examiner.

In social science fields, students must build a framework for discussion from the reconstructed literature, as Bazerman (1988) has explained. Careful alignments with established theory and methods must be expressed and justified, as we have seen.

There are complex levels of tacit learning involved in building an argument resting on existing literature, given the skilful art of persuasion in scholarly texts identified by Gilbert (1977) and Small (1978). Few of the persuasive tactics are boldly obvious in scholarly publications, but mastery of them is nonetheless an indication of disciplinary expertise.

The conventions for appraising existing research are, in general, reasonably polite in science fields, where discrete judgements are mainly executed by selection or omission rather than by value judgements, as Becher (1987b, 1989a) observed.

There is evidence, too, that citation practices uphold discrete norms such as paying homage to exemplars (Weinstock, 1971) and being persuasive in discipline-appropriate ways, as suggested by Becher (1989a) and Parry (1998). Discrete norms relating to disciplinary etiquette in science exist as well. A few students from experimental science fields such as psychology and geology, for example, espoused disciplinary conventions regarding the importance of publishing, while hinting that, in realizing this objective, their supervisors were capitalising on their efforts. Science students are sometimes concerned about being expected to do the bench work of their supervisors, a situation they feel it necessary to accept in order to 'get on'. There are many unspoken rules about how to get on in the hierarchical communities of science, mastery of which is reflected in the capacity of most students to compromise their short-term needs or gains for the longer-term achievement of acceptance in the research environment and in the wider discipline.

Disciplinary etiquette among specialized knowledge communities, like their knowledge bases, constantly evolves and changes as new influences become accepted. This means that the kinds of skills that students must master, particularly in transdisciplinary fields, also evolve and change. The degree of complexity and uncertainty faced by students under these circumstances grows more and more challenging. Can it be any wonder that, at least in the United Kingdom and Australia, graduate skills are increasingly an issue, and especially so for employers?

If doctoral students can learn to put aside their short-term interests to promote longer-term goals; if they can learn complex counter norms as we have seen in abundance; if they can learn to discern when to ask a question, as Gerholm described, then they can learn which conditions constrain the use of knowledge for different purposes, different audiences and different situations. And they can apply them accordingly.

7.2. Relevant Study Programs

The disciplinary core may well include an eclectic grab-bag of opportunistically-collected knowledge, as Becher and Parry (2005) argue, but at the centre of any eclectic mix is the powerful core of disciplinary knowledge. The disciplinary core provides credibility and reliability it to more socially-distributed forms of knowledge (Nowotny, Scott and Gibbons, 2001) – with all its inherent uncertainties. Attempts to prescribe skills training to address these uncertainties are likely to achieve mixed success. A strong disciplinary grounding, however, provides the platform for crossing boundaries into other knowledge fields and cultures. So it is a disciplined background that is the essential starting point for novice researchers in managing uncertainty, boundary changes and the closer relationships between knowledge makers and knowledge users that are associated with new forms of knowledge and knowledge production.

The diminution of the doctoral research enterprise by reducing it to skills training is amply illustrated by the comments of a professor of agriculture: "Lets' face it. . . . . You can't standardise things that aren't standard. . . . They try to regularise things that are part of a process of discovery, and you kill the velvet touch if you list everything out and make people follow form and routine." The message here is clear: much of what is learned is unconsciously committed to memory by tacit means when the information is relevant and timely. Readiness and receptiveness to learn are not qualities that can be ordained by enforcing routine learning tasks – especially when they bear no relevance to the research at hand – upon doctoral students.

Gerholm's (1990: 268) argument that: "The more frequent the contact between experienced researchers and their students, the greater the likelihood that the tacit knowledge of the discipline is being passed on" still resonates strongly. His argument signals the importance of tacit learning even though it does not account for the vastly different cultural settings of different fields of knowledge, nor does it account for the degree of change and dynamism in the contemporary nature of knowledge.

The disappointing completion rates and long completion times for students in social sciences and humanities fields now run the risk of extending further into transdisciplinary knowledge fields. Whitley (1980: 311) observed that disciplinary norms and procedures appear to be less clearly formalised in fields where uncertainty is a feature, because the means by which they may be pursued are also less obvious. Whitley argues that this feature explains why PhD research is less highly organised and structured in these settings.

It is worthwhile at this point to reconsider the importance to doctoral students of making a set of commitments to a disciplinary core – and values, conventions and traditions associated with it. There can be little doubt that making these commitments may take longer, and be a more complex process, in interpretive fields of knowledge and in those areas where uncertainty is a feature. Against this background governments attempt to reduce funding for doctoral candidature to a formula that is applied across the spectrum of fields of knowledge. Without financial support from new stakeholders in higher education, newer and more contemporary fields of knowledge cannot easily be supported. But, as knowledge is distributed more widely in society, its reliability will depend upon the strength of the commitments made by knowledge makers – including doctoral students – to core values.

CHAPTER 9

IMPROVING THE DOCTORAL EXPERIENCE

1. CONTINUING ISSUES IN DOCTORAL STUDY

Having acknowledged the complex and sophisticated nature of learning at the level of the doctorate, a number of issues concerning the relevance, appropriateness, quality and diversity of doctoral study programs remain. What should be made of the contemporary nature of research – and of knowledge production – in reconceptualising the notion of doctoral study?

1.1. The Nature of the Award

These questions address the need, more pressing than ever before, for reconceptualising doctoral study programs and the doctoral award itself. How can institutions support doctoral programs that are closely attuned to the intellectual and professional communities that, in addition to the students themselves, have a stake in their outcomes? The market demand for credentials has to be acknowledged, but to date the doctorate signifies a wide range of academic achievements with an assumed equivalence of standards. The third wave of the Bologna Process has raised the spectre that this assumption is not robust. The student market increasingly requires doctoral awards that can be articulated across different systems and countries. As a result, national systems must demonstrate the equivalence of their doctoral awards with those of other countries. Once European Union countries have reached agreement through benchmarking processes, it will be difficult to market doctoral awards that are not demonstrably equivalent, or do not have the same cache as those offered elsewhere.

Higher education systems are now faced with some difficult questions and associated challenges. Should a doctorate represent one kind of intellectual achievement or different kinds of intellectual achievement? Should there be reliable standards that the title 'doctor' represents? If so, how might standards be determined and different kinds of awards distinguished? How can, or should, institutions support the largely tacit nature of learning at the level of the research doctorate? Who should accept responsibility for improving the doctoral experience and making it more relevant: individual students: supervisors, departments, graduate schools, universities or governments? These are complex questions for which there are no simple answers; accordingly they are the subject of this final chapter.

1.2. Satisfaction and Relevance

A study by Harman (2002) found that only 56% of Australian PhD students were satisfied with their experience of research supervision, a finding that does not reflect well on attempts over two decades by governments, systems and institutions to improve the quality of the doctoral experience. Harman argues that those who were satisfied were more likely to be working in collective research cooperatives in applied science fields. Harman's findings are a worthy pointer to the satisfaction among collective, applied fields of science with established, normative research cultures. However, the findings also point to considerable dissatisfaction with research supervision outside this narrow definition, which of course includes research in the more individualistic fields of the humanities and social sciences. Indeed there are implications for transdisciplinary and applied research fields to say nothing of the applied professions – all areas whose profile and importance to the community generally is increasing, as many researchers have observed.

On the one hand, there is currently a need to design doctoral study programs with the needs of a broader range of stakeholders, including career professionals, industry, commerce, community groups and government in mind. This extended broader group of stakeholders has particular expectations about the skills and capacities that study programs should foster in graduates. On the other hand, there is the desire of governments to make doctoral study relevant, effective and cost efficient. Efficiencies are often achieved by concentrating the research enterprise into key areas and centres where doctoral students are co-located in teams mimicking those of the natural sciences. Whether these enclaves fostering doctoral research are improving completion rates and times in non-science fields is yet to be demonstrated, however. In this context we have seen how the doctoral experience is quite naturally a longer process, for epistemological reasons, in individualistic fields of study. Should candidature times in these fields be limited? If scholarships and stipends for doctoral study are constrained by the norms in science, as they are in some national systems, the implications for the maintenance of standards will inevitably come into question.

To begin to address these questions, the nature and purposes of contemporary doctoral study need to be clarified. It is also necessary to draw together some features of doctoral study that do or should transcend national boundaries, and to speculate a little on where developments seem to be heading.

1.3. Social Settings for Learning

Henkel (2000) has identified a central issue. It concerns whether it is possible or indeed desirable to apply to other disciplines those opportunities for learning disciplinary features, explicit and inexplicit, that have been found to prevail relatively successfully in the natural sciences. In the preceding chapters, we have noted the marked differentiation in disciplinary values, aspirations and socialising opportunities in which doctoral students are immersed. Particularly striking is the growing awareness among doctoral students of the need to acquire a certain disciplinary savvy whose social and epistemological elements are highly specific to their specialisms. Yet Becher and Trowler (2001: 21) observe how postgraduate education, along with other areas of academic endeavour, have largely been homogenised to suit managerial imperatives:

In most cases, the identified contrasts between established cultural assumptions and practices may be overridden by uniform, undifferentiated policy requirements. In consequence such requirements can result in anomalous and insensitive impositions, which are liable to be tacitly, if not overtly, rejected by those called upon to adopt them.

A case in point concerns the well-documented sense of isolation expressed by doctoral students, which may be both social and intellectual. Delamont, Atkinson and Parry (2000: 177) acknowledge that the social relations in the laboratory or research group can mitigate isolation. But it needs to be recognised that, in the more individualised fields of the social sciences and the humanities, social interaction is not an end in itself because collaborative outputs are not always appropriate. In these fields, social interaction is a means to opportunistic learning, intellectual stimulation and validation. It therefore comes as no surprise that, after the recent policy and structural developments in doctoral programs in the UK (noted by Henkel, 2000) and in Australia (noted by Neumann, 2002), improvements in completion rates and times are yet to be reported.

A related issue concerns the extent to which it is possible or even desirable to articulate the range of disciplinary norms that shape the making and reporting of knowledge in specialised knowledge fields. Since, as we have observed in earlier chapters, disciplines share core features but in essence are dynamic, constantly evolving, splintering and growing, to attempt to make concrete all aspects of their fabric would be to constrain their capacity for innovation, development and change. At the same time, the greater contextualisation of knowledge making in society brings with it the need for social savvy that extends well beyond the boundaries of scholarly communities, into society and communities of practice.

Science and scientists now face an *agora* with multiple publics and plural institutions, such as the mass media, which vigorously conduct their own negotiations. They are faced with a complex bureaucratic and administrative web of funding agencies that devise their own policy goals, guidelines, assessment procedures and allocation mechanisms. Researchers have had to develop a new range of skills in communicating with their potential funders, writing grant proposals and promising enticing outcomes that cannot really be specified in advance. They face an industry-business landscape which itself is the object of radical restructuring (Nowotny, Scott and Gibbons, 2001: 206).

Juxtaposing the authority of epistemology with the need for relevance, Nowotny, Scott and Gibbons (2001: 198) propose a 'third way' for science that fits with a reconceptualisation of doctoral study: "A more nuanced and more sociologically sensitive epistemology is needed which incorporates the 'soft' individual, social and cultural visions of science as well as the 'hard' body of its knowledge". With the impending establishment of the European higher education area by 2010, requiring equivalence among doctoral awards across European national systems, such a reconceptualisation is pressing. Can doctoral programs be customised to meet community expectations and yet retain equivalence – and standards – across national systems?

As the global economy imposes itself further on higher education, the need to benchmark standards in university awards – including the doctorate – looms larger and larger. Neumann (2002: 175) highlights the urgent need for policy to encourage and direct diversity in doctoral education. She draws together Rip's (2000) view that the traditional PhD is sufficiently robust in scope to accommodate the diversity of new knowledge with Brine's (2000) assertion that Mode 1 and Mode 2 knowledge has co-existed comfortably for some time. Neumann (2002) acknowledges that Mode 2 knowledge, characterised by diversity, is essential to the capacity to perceive issues from a number of perspectives; she also acknowledges that perspectives belong to particular knowledge domains.

The debate about relevance in doctoral programs is bringing to the attention of governments the need to benchmark standards and the need to determine the equivalence of awards that earn the title 'doctor'. While the debate simmers relatively slowly among policy researchers outside Europe, universities within and across national systems are conferring doctorates for very different kinds of knowledge achievements. The conferral of diverse doctoral awards is a response to demonstrated market demand for doctoral credentials of many different kinds. This outcome seems to go largely unremarked in some national systems, perhaps because, as Usher (2002:152) puts it, "If we in universities are to encourage a diversity of doctorates we need to know what we might be losing as well as what we might be gaining". In the meantime, the equivalence of different kinds of awards is largely taken for granted within the community, though academics themselves would argue this point.

The need for diversity in doctoral awards and study programmes derives from the growing community understanding that new kinds of knowledge societies are emerging, altering an already changing higher education landscape. Becher and Trowler (2001: 15), for example, point to seismic systemic and institutional changes in higher education systems in recent years. They note the explosion of knowledge and point to constant growth, broadening, diversification and fragmentation of disciplines, observing the dignification of some and the decline of others. They also acknowledge the influence and impact of market imperatives and student instrumentalism.

Delanty (2001: 54), in a similar vein, notes " . . . . the growing influence of professional training (seen). . as a response to the demands of the economic system, which creates the need for a public system of accreditation". The managerialist stamp upon doctoral programs in the UK and Australia particularly has seen academic departments come to accept that doctoral students, like all students, are paying customers who seek value for money, which is accounted for by fitness for purpose and by awarding credentials.

The result is that the market for a doctoral credential is growing, with a diverse range of programs being developed in response. Scant attention, though, is being brought to bear on the responsibilities of individual institutions that, for the most part, accredit their own awards. As the global market pressure increases for awards that provide relevance, compatibility, and mobility, national systems including those outside continental Europe, will need to be assured of their equivalence and of the achievement that the title 'doctor' represents. The reputational value of awards and their conferring institutions will become increasingly predominant as national systems position themselves to demonstrate equivalence. One outcome of these developments is that the dialogue about doctoral education is moving from a focus on the individual student or the individual supervisor to institutional and systemic challenges that require policy-related solutions.

1.4. Standard Setting

An obstacle to effective benchmarking of standards at the level of the doctorate is that universities, after all, are self-accrediting institutions in the majority of countries. The proliferation of the different kinds of doctoral awards documented by Usher (2002) – traditional PhDs, coursework doctorates, professional doctorates, doctorates by publication and by project – is evidence of the extent of diversification overall, and of course there is variation among institutions within national systems as well as across national boundaries. The difficulties in recognising equivalence in continental Europe as signalled in the third wave of the Bologna Process (Bergen 2003) are not restricted to that continent.

For many years the different standards expected from different kinds of doctoral awards among higher education systems have gone largely unremarked. The self-accreditation and self-rubrication of standards is a time-honoured tradition with a pivotal place in university life. It has traditionally cemented the role and function of the university in society, and so it is a value that few in academe would see compromised. And yet, the time has come when the role and functions of the university as well as their awards, are being reconceptualised by society at large. With the proliferation of doctoral awards on offer, equivalence may no longer be taken for granted. Until the tension between market demand for the doctoral credential and undifferentiated standards for fundamentally different kinds of awards is resolved by national systems, a range of organisational problems with doctoral study and supervision cannot effectively be addressed.

1.5. Robust, Contemporary Research Training

The literature on doctoral education over the past two decades covers organisational and intellectual aspects of research training, supervisory responsibilities, the typical causes of poor completion rates and times. In much of this literature, there seems to be an underlying assumption that doctoral programs are equivalent, at least in terms of their representational value. Little distinction is made between the different kinds of purposes that doctoral programs serve; those distinctions that do exist are reputational and are largely quarantined within national systems. However, the growing demand for doctoral awards also presses for national systems to assure standards and demonstrate equivalence.

For the undergraduate curriculum, Biggs (1999) argues, as he has for almost two decades, that the focus of the curriculum, its teaching and its assessment, should be upon the desired *learning outcomes*. To date the emphasis in the literature on doctoral study has been concerned with the *organisational processes* of doctoral study rather than with the intended learning outcomes.

The kinds of learning outcomes envisaged by students and other stakeholders in doctoral study outcomes – their communities of practice – now assume more importance than previously. This is not simply because knowledge is more socially distributed, nor is it that students are more instrumental or vocational: it is because students have highly differentiated learning outcomes in mind. No longer is the PhD simply a licence to practice as a university researcher and teacher. There are multiple kinds of learning outcomes envisaged for the award by the various kinds of stakeholders concerned. From the students' point of view, doctoral study programs are becoming more individualised and in many cases more applied, in keeping with the contemporary nature of Mode 2 knowledge.

At the same time, doctoral study for individuals has undoubtedly become more complex. In turn, the capacity to identify with a core disciplinary area while still being able to navigate the appropriate related knowledge bases and knowledge cultures becomes more difficult. It is certainly problematic for students, and their supervisors for that matter, to come to terms with fast-moving conceptual and epistemological developments in their fields of knowledge. Undoubtedly the growth in Mode 2 knowledge will continue and mature, with market demand for individualised doctoral programmes flourishing as communities of practice grow and develop. The requirement for equivalence to be demonstrated among similar awards requires a reconceptualisation of doctoral programmes that takes into account their different aims and achievements.

1.6. Changing Demand for Doctorates

Tinkler and Jackson (2000) note the different perceptions of standards at the doctoral level, and like many recent observers, note also the increasingly instrumental nature of student demand for the award. Bourner, Bowden and Laing (2001) also acknowledges these diverse perceptions as well as the diversity of awards on offer. Bourner points to the time-honoured external examination process as being central to the maintenance and parity of standards for the award. In the light of the growing pressure in many countries for doctoral programs to be more relevant and efficient in meeting the needs of higher education customers, the different kinds of awards and the standards they represent deserve further consideration.

Bourner and Laing (2000) distinguish between professional doctorates as awards for professional people who may already have a well-established career, but who wish to be able to undertake research in their field, and the PhD, which remains essentially a research-training award for professional researchers in academe. This distinction is useful because it implicitly recognises the vocational aspirations of the student market and of their need for credentials as a licence to practice in different kinds of professions and with different kinds of communities. Apparently national systems, and their governments, are less prepared to make this distinction. Neumann's (2002) evaluation of the Australian system highlights the conflation of research awards, coursework awards and professional awards as an artefact of a research-primed funding platform for doctoral awards. Other national systems have difficulty distinguishing different kinds of awards for other reasons, such as the more politicised French system that places a higher value on the prestige of the awarding institution than on the award or doctoral title itself (Neave, personal communication 2003). To date, by way of contrast, the Norwegian system awards only PhDs. While post-masters qualifications are on offer in professional fields of practice, these do not earn the title 'doctor'.

Yet other contextual and historical factors have given rise to the different conceptions of the doctoral degree. The title doctor is awarded for different standards of research and coursework across different national systems. Some observers (Rothblatt, 1997; Delanty, 2001) consider that one impetus for research degrees becoming more applied and transdisciplinary, and therefore closer to their communities, may be derived from the US system. While this is arguably the case, there is no doubt that the social and economic needs of various national systems have played a part in shaping the nature of their doctoral programs (see, for example, Clark, 1993). The US doctorate, for example, while essentially a research award, includes substantial components of coursework By way of contrast, doctoral programs in former eastern European countries differ markedly. Unlike their western counterparts, which are ". . . seen as transmitting socially useful and productive knowledge (with) . . . market factors . . . (gaining) . . . ground in institutional management", former eastern block countries adopted the Soviet system and quarantined research degrees in academies and more applied research into professional colleges (Rozsnyai, 2003).

Developments such as these are a foil to the largely external system of doctoral examination that safeguards the standard of the PhD. But there are subtle nuances to be noted among doctoral standards and in the processes of examination, too. One supervisor, commenting on variations in the standards and *types* of doctoral theses he had supervised said: "(Hers) was a very applied thesis. She had to do a lot of skills development – she had to learn cluster analysis in the end – but it wasn't a theory-laden piece of work. Yes, I did scratch my head, and in the end I chose an examiner from Berkley and two from the UK. The Berkeley examiner was my insurance. . . . " There is an underlying assumption that standards are reputationally derived, which is not always the case. As Mode 2 knowledge proliferates, the reputation of a single distinguished exemplar may not be sufficient to assure the standard of the work as socially robust.

Within national systems, there is a tacit understanding of the quality of a doctoral award from the different institutions – explicitly so in France. Added to this, there is a tacit understanding that different types of doctoral awards serve different kinds of purposes. Some are directed more towards a research career in academe while others are directed towards particular professions, the Doctor of Business Administration and the Educational Doctorate being illustrative examples. Recognition of these important differences is yet to be made, making attempts to articulate equivalence across national borders difficult indeed. Added to the lack of recognition is the recent trend towards increasing coursework components in an effort to make the PhD a more versatile, vocational award.

One way to reduce confusion is to identify the core components of a *philosophical doctorate*. Coursework undertaken towards developing skills and capacities directed towards multiple other publics and purposes would then be quarantined in other awards. The issue at stake is that the cache of the doctoral award is changing; multiple publics and purposes require more than a sophisticated knowledge base; they require that knowledge to be socially distributed. There is the danger, however, of decontextualising doctoral research and distancing it from its agora, so there is a need to forge stronger links between institutions and their communities of practice. This approach is one being adopted by default in some national systems, where the PhD is no longer a licence to teach; a tertiary teaching award is required as well.

Another approach would be to broaden notions of the doctorate to accommodate whichever coursework and skills training are deemed appropriate by individual institutions. In this approach, institutions lose their capacity for self-accreditation, with awards being judged for robustness in their agora. Benchmarks and parameters would need to be set by governments, experts and knowledge communities in tandem, the danger being the potential for a diminution of the epistemic foundations of doctoral study.

1.7. Thesis Examination

Thesis examination has been the focus of several studies in which examiners' reports have been reviewed (Pitkethly and Prosser, 1995; Ballard, 1996; Johnston, 1997), but Delamont, Atkinson and Parry (2000: 41) have concluded from interviewing experienced examiners that "the real role of the examiner is to judge whether the student has mastered appropriate indeterminate skills and displayed the right indeterminate qualities".

Mullins and Kiley (2002) in a more recent Australian study found that the major influences on examiners concerned institutional criteria for judgement, a tacit understanding of standards appropriate to the field, the quality of the student's existing publications in the field, and their knowledge of the standing of the student's supervisor and department. All of these influences point directly to the need for examiners to possess a sound understanding of the core field of knowledge, including its major methodological and theoretical developments, as a basis for making judgements about appropriate standards, skills and qualities.

The typical processes for thesis examination vary considerably among national systems of higher education. Not only is this problematic in terms of the standard of the award, but it is also problematic in terms of the rigour or trustworthiness of awards. Until now, society has assumed that disciplinary cultures assure the standard of doctoral research through the largely external processes for thesis examination. With the proliferation of awards and greater accessibility to candidature in developed countries, however, doctoral awards have become a marketable commodity. Assuring the quality of doctoral awards has never been more urgent. A recent UK code of practice (Quality Assurance Agency for Higher Education, 2004: 24) has in part addressed this problem by advocating that institutions ensure that examiners submit separate, independent written reports before the viva and a joint report after it, and that they identify ways of handling cases where examiners are unable to reach a consensus view of the outcome. There remains, though, a need for national systems generally to address issues associated with thesis examination, including the expected standards represented by different awards.

The time-honoured assumption that disciplinary enclaves provide systematic training and codes of practice for disciplinary values and traditions is not robust enough to account for the more individualised, applied forms of Mode 2 knowledge now being produced at doctoral level. Benchmarking of examination processes to make them more systematic and socially reliable would considerably assist national systems, and individual institutions, to assure and demonstrate the equivalence signalled by the Bologna Process. The UK code of practice (Quality Assurance Agency for Higher Education, 2004: 24) goes some distance in this direction by calling for assessment criteria and individual, followed by joint examiners' reports, as a means to making examination more systematic across institutions. However it stops short of calling for more systematic institutional policies and practices, such as requiring more than two examiners or calling for assessment panels. While the code identifies the need for prepared and well-qualified students who are inducted to postgraduate research cultures, it is silent about the need for prepared and well-qualified examiners who are inducted to field-specific standards and assessment practices!

2. IMPLICATIONS FOR INSTITUTIONS

In an in-depth study of a large research university, Cullen and colleagues (1994) indicated that institution-wide organisational units for supporting research degree candidature would address student dissatisfaction with intellectual isolation, inadequate resources and inappropriate supervisory practices, particularly in settings where research is novel or transdisciplinary. They postulated that such organisational units might offset the protracted completion times and poor completion rates especially prevalent in the humanities and social sciences. At that time, the inadequate allocation of resources to doctoral research in some academic departments, the humanities in particular, was widely reported.

2.1. Administrative Convenience

There seems to be little doubt, based on the willingness of institutions to take up this idea in both the United Kingdom and Australia, that administrative efficiencies accrue to institutions with such organisational units. It is reasonable, too, that students in individualistic fields of study are better off socialising with a disparate group of researchers than having no reference group at all. But these new organisational clusters obscure the role of socialisation and induction to disciplinary values, traditions and conventions. We have seen that both students and their supervisors are acutely aware of the need to develop disciplinary savvy that carries a student onto the international stage and to multiple publics, so there are bound to be mixed messages when the integrity of disciplinary values and conventions from across the range are homogenised towards those which successfully prevail in the natural sciences.

A recurring issue in the related literature has concerned resource allocation to doctoral research. The tightening of funding in many systems has provided a logical impetus for the establishment of graduate colleges and the like. Resource allocation typically varies across academic departments due to the pace and cost of the research enterprise, but there is room for institutional policy on resource allocation that acknowledges both the knowledge-related and socially related characteristics of learning in disciplinary settings. In many science departments, these inexplicit conventions and practices are part of a long tradition. In other disciplines, the parameters may be inexplicit, constantly changing and maintained by scholars whose demography is sparsely populated and widely scattered: it generally takes doctoral students a long time to work them out.

2.2. Setting Parameters

For some time it has been recognised that institutions have a responsibility to ensure that codes of practice for doctoral supervisors and their students are implemented. But to date these have largely been concerned with matters in the administration of candidature, rather than with facilitating the induction of individuals to international and cosmopolitan knowledge cultures. While there will necessarily be differentiation in the expectations of supervisors and students across disciplinary and departmental settings, there is still a need for broad parameters and benchmarks to be articulated. A frequently recurring question from both students and supervisors alike in this investigation was: "am I doing about the same kinds of things as everyone else (in my area)?" Supervisors want practical advice on supervisory strategies, but they also want to know how much responsibility they should take for the development of knowledge-related and socially-related disciplinary know-how in their students.

Students also need to know how much they can expect their study programme to be customised to suit their individual needs. They need to know how much they can rely upon their supervisors' efforts, and how much they ought to use their own initiative. The list of individual needs identified by students in this investigation, about which institutions need to set parameters, is extensive. How much and what kind of assistance with writing should be provided? How much responsibility for covering an adequate substantive literature? Does the supervisor have to endorse submission of the thesis to examiners or not? How much conference and travel assistance should be provided annually? Should the student be wholly responsible for data analysis or will there be assistance? If assistance is available, what kind is it and how much of it is appropriate? Is there assistance with the writing of English? If so, how much and what kind? How do contributions to intellectual property work – what weight does the contribution of ideas have compared with the contribution of writing up? Who takes ultimate responsibility for the choice of methodology and the effectiveness of its implementation? What kinds of probationary conditions are there on candidature? Is it appropriate for doctorates in certain fields of knowledge to take much longer to achieve than others? What are the requisite minimum qualifications and experience of supervisors?

Added to explicit parameters for the above agenda, guidelines for students need to take account of the nature of the learning that takes place at the doctoral level. Institutions or departments are not required to set parameters for doctoral students' socialising opportunities, and yet these are pivotal to success in every field of study. New students cannot possibly know how to make contact with established scholars, and they need to participate in global dialogue and critique from the beginning. The available socialising opportunities for particular fields of knowledge should be identified, together with an indication of the extent to which the institution will financially and practically support engagement with them. Students have to learn how to make informed judgements about "how to get on" and what to prioritise.

2.3. Choosing Study Programmes

A perhaps unpalatable institutional responsibility, especially where doctoral programs are marketed to students from other countries, concerns the explication of different kinds of doctoral degrees and the markets to which they are suited, together with the standards and examination requirements associated with different forms of the degree. There is already a reasonable equivalence among undergraduate awards within fields of study across national systems – physicists, historians, veterinarians all share roughly equivalent undergraduate education in developed countries. This may also be the case among different kinds of doctoral awards – professional doctorates in particular – but equivalence across professional and coursework doctorates and the PhD is not guaranteed. McWilliam an colleagues (2002: ix), surveying those on offer in Australian universities, found "131 programs . . . in 35 of the 38 Australian public universities, exhibit(ing) a wide range of structures and features". They noticed that some programs exhibited 'deep' levels of linkage with professional and industry bodies, and recommended this kind of linkage as developing a community of learning that involves doctoral researchers with community, industry or business stakeholders. They failed, though, to acknowledge the need for equivalence in or the need to explicate standards among these diverse groups of stakeholders in their policy recommendations. The need remains for agreed standards, regardless of the field of study, to be explicated because not only are there different kinds of awards, but the long-acknowledged philosophical doctorate continues to command the greatest cache in so many national systems. While institutions neglect to articulate these differences in their course offerings, they fuel the creeping credentialism so often noted by policy researchers.

3. IMPLICATIONS FOR SUPERVISORS

Differences in the nature of doctoral awards and trends in the nature of the research enterprise itself are also pertinent concerns at the level of the faculty or department. As research becomes increasingly applied and transdisciplinary, supervisors can no longer be expected to be authoritative on all aspects of doctoral research under their management. Not only do supervisors need sanctions to be able to articulate their limits to competence in broad disciplinary groupings, but they also need protocols for drawing upon supplementary expertise as appropriate, particularly when the required expertise is from outside the institution.

Part of supervisory responsibility is to help students to learn disciplinary lore. But it is also evident that doctoral students must use their own initiative in the learning process, learning to seek out for themselves appropriate opportunities for learning socially-related and knowledge-related conventions. One supervisor said of his student: " . . . . after all, it is their thesis. I shouldn't have to be responsible for (the final quality of) it." But the supervisor is responsible for facilitating the socially-based learning opportunities that underpin it.

A key responsibility of academic departments is to clarify and market the various kinds of doctoral awards on offer. While the cache of the title 'doctor' may be the most important feature for some, other students are concerned with different kinds of learning outcomes. Some are vocationally oriented, some are motivated by a burning question, some are updating their qualifications and some are honing their research skills in an area of expertise. At present, distinctions among awards are clear-cut in France, where there are no assumptions about standards and instead the reputation of the institution signals the standard of the award. But in many other systems, standards are taken for granted and awards are marketed accordingly.

It is clear that connection and involvement with knowledge networks is essential if supervisors are to have any confidence in guiding students towards appropriate learning opportunities. To achieve this minimum requirement, departments should recognise and support active supervisors, setting minimum standards of "connectivity" and scholarship as a requirement for practice. At the same time, there has long been a need to identify and support novice supervisors through induction and training programs and through mentoring programmes. There remain to be addressed within departments many long-standing questions such as: what are the minimum qualifications, experience and expertise of supervisors in departments? What training is available, and what are the intended outcomes of the training? Whether these kinds of policies and procedures should be invoked at the institutional level or at the departmental level is open to debate, but individual supervisors should not be left to second-guess the nature of their role in supporting the intellectual development of their students.

The induction of supervisors is a much-discussed issue, as we have seen. Surprisingly little discussion, though, engages with the need for inducting examiners to assessment practices and the articulation of standards. For the undergraduate curriculum, Biggs (1999) adopts a constructivist approach to demonstrate how intrinsically linked learning objectives are to what is assessed; the identification of standards is at the heart of the learning enterprise and the assessment enterprise. For research degrees, and especially at the level of the doctorate, the articulation of standards is far more complex, involving as it does the development of disciplinary savvy. Doctoral students find the identification of standards to be at the heart of the learning enterprise, but there is scant assurance at present that the identification of standards is at the heart of the assessment enterprise.

4. IMPLICATIONS FOR ACADEMIC DEPARTMENTS

Delamont, Atkinson and Parry (2000: 153) showed that, socialisation in the experimental sciences is distinctly different from socialisation in the interpretive sciences. They distinguished between socialisation in the immediate research environment of academic departments in science and interpretive fields. In experimental science, which is hierarchical and overtly structured, socialisation tends to be positional, with clear lines of social control. Alternatively, in interpretive fields of knowledge, it is more likely to be personal, social control based more upon social relations encompassing coercion and persuasion. These researchers consider that students' accounts of academic identity are polarised – "from the position of underling to that of professional colleague."

The distinction between positional and personal socialising cultures is useful in underscoring the complexity of disciplinary identity. The marked differentiation between the natural sciences and the social sciences and humanities points to the role of socially-related aspects of knowledge, involving underlying values and epistemological foundations, forms of social engagement and control, aspirations and forms of discourse. It also echoes Knorr-Cetina's (1999) distinctions between open and closed cultures governing the social reproduction of knowledge in science. Delamont, Atkinson and Parry (2000: 1 55) explain:

The social and intellectual identities of the science and social science PhD students are different. The social organisation of knowledge production and reproduction constructs different selves for academics and for doctoral students. The differences in social and intellectual roles make the everyday life of the graduate student very different.

Departments that are large and reasonably homogenous may be able to be sanguine about the complexities of developing disciplinary identities. But among departments, and other kinds of organised research units, whose populations are epistemologically diverse or heterogenous, two considerations in particular are important. The first concerns the cosmopolitan nature of academic disciplines, whose cultures are not bound by the organisational role of departments because they are internationally oriented. It is through the opportunities for socialisation in the broader scholarly milieu that departments can provide doctoral students with benchmarks for developing disciplinary identity and savvy. While established scholars need affirmation and feedback on their research-in-progress, so too do doctoral students. Affirmation and feedback need to be both timely and relevant to individual students, and it is at the level of the department or research unit that these needs can be planned for and strategically met.

The second concerns the very different disciplinary cultures that can be represented in a given department. Diverse knowledge bases and theoretical perspectives may co-exist, so it is especially important that doctoral students not be expected to comply with policies and practices for skill development and other kinds of learning that are homogenised for administrative convenience. Delamont, Atkinson and Parry (2000: 15, and 10) acknowledge the cultural difference between departments and disciplines but argue that "most student effort must go into learning the ropes of the local organisational routines and requirements of their departments." Yet we know from Henkel's (2000) study the importance of the discipline in the development of academic identity. The discipline is also the primary source for doctoral students, as studies by Becher, Henkel and Kogan (1994) and Parry and Hayden (1994) have demonstrated. The implications for departments are clear: socialisation in the department cannot and should not be expected to take the place of induction to an international community of scholars.

5. IMPLICATIONS FOR DOCTORAL STUDENTS

Early researchers (Rudd, 1975; Welsh, 1981, 1982; Moses, 1984) noted that it has long been an expectation that doctoral students should use their own initiative in the learning process. Not much has changed in these terms in the intervening decades. But the knowledge landscape has changed considerably, as have the cultures inhabiting it. And so it is timely to reflect on the nature of the learning involved in doctoral programmes.

Fields of knowledge have evolved to take more account of the communities they serve. Students have become more vocationally oriented; more supervisors than ever before are faced with supervising in novel, transdisciplinary and increasingly applied research areas. Institutions have tried to improve student satisfaction with and the efficiency of doctoral programs – along with completion rates and times. Institutions have also become more entrepreneurial, establishing different kinds of doctoral awards to meet market demand. In addition, governments have permitted large increases in the numbers of doctoral students admitted to programs.

These are giant changes for supervisors and doctoral students alike. As knowledge becomes increasingly transdisciplinary and individualized, the complexities and uncertainty associated with making knowledge must seem daunting to aspiring students and novice supervisors.

For individuals enrolling in doctoral candidature, all the issues raised here are relevant considerations in what is, after all, a significant intellectual achievement. In considering them, and in considering the findings of the study reported here, it is worth recapitulating on the role of knowledge cultures in shaping the ways in which knowledge is made, as well as the ways in which it is reported. After all, disciplinarity lies at the heart of the doctoral experience: understanding its nuances and epistemological significance are needed to succeed.

Academic disciplines are cognate communities, while universities are organisational communities. It is in organisations that systematic training and development programs are provided for doctoral students and supervisors. It is also within these organizations that examiners ought to be able to access systematic training and development from those with demonstrated expertise. And yet these training and development programs have a cosmopolitan dimension that is ultimately more powerful than the organisational setting. It is therefore essential to keep in mind that the success of the doctoral thesis depends upon the appraisal and recommendations of the examiners concerned. Their judgement is subjective and conditioned by international frames of reference that are readily observable in the relevant scholarly literature.

Opportunities to develop disciplinary savvy are to be drawn from the international arena, the national stage, the institution and its research arms, the academic department or research centre and the scholars who inhabit these domains, as well as from supervisors themselves. Grounded savvy in the discipline is shaped by a range of significant others who are actively engaged in, and intellectually marshalled by, their disciplines. Within these, cosmopolitan, international research networks that increasingly include stakeholders in the associated community of practice have the abiding influence and normative power. Social savvy may well be gleaned largely from the immediate research department, research group or graduate school, but cultural savvy and discourse savvy must be developed in the context of the broad disciplinary arena.

6. AND FINALLY. . . .

In the development of effective doctoral programs and in the kinds of experiential learning opportunities they might comprise, there is a pivotal consideration. It derives from the distinction drawn in undergraduate teaching and learning by Neumann, Parry and Becher (2002), where knowledge-related aspects of disciplinarity and socially-related ones are differentiated, but both are given equivalent epistemological importance. Because opportunities for learning both knowledge-related aspects as well as those that are socially bound are essential, opportunities for practising both need to be explicitly provided within doctoral programs.

Doctoral study is a cultural phenomenon; it is a rite of passage in the disciplinary sense. Successful completion of the doctoral thesis reflects mastery of distinctive values and norms that shape the expression of knowledge in particular disciplinary fields. For a doctoral program to be effective, it must address several socialising considerations. The first is that a range of individuals play a part in inducting a doctoral student to a particular disciplinary culture; principal among these may be the supervisor, though frequently, other individuals – such as those participating in the broader community – play crucial roles. There is a need for the relationships between disciplines and communities of practice, which are so fused and yet not the same, to be acknowledged and better understood.

There is a need for a systematic approach to the identification of intended learning and knowledge achievement outcomes associated with different kinds of doctoral study programmes. For doctoral awards to maintain their robustness and reliability in society at large, the standards that are aligned with particular outcomes must be widely understood.

In addition to the articulation of objectives and standards, there is a need for the socially-based dimensions of learning to be better understood. A thesis that is successfully completed represents effective induction to a field of knowledge, yet the role of the broader range of stakeholders in the field is generally given little importance unless the doctoral research is stakeholder-funded. Indeed, there are no benchmarks for determining what would constitute a reasonable or sufficient range of socialising opportunities for induction to take place. Against this background, it has already been noted that doctoral students may fall into the categories of 'cue seeking', 'cue conscious' and 'cue deaf'. Simply raising awareness of the norms that shape the expression of knowledge would be of great assistance to supervisors and students alike. Nowotny, Scott and Gibbons (2001: 262) reach a similar conclusion: "Just as publish or perish is underpinned by certain rules of the game to which scientists and their peers have agreed to adhere, so the opening up of science towards the agora presupposes and necessitates 'rules' of a game that partly still wait to be established."

Amann K, Knorr-Cetina K (1989) Thinking through talk: An ethnographic study of a molecular biology laboratory. In: Lowell H, Jones RA, Pickering A (eds) Knowledge and society: Studies in the sociology of science past and present. London: Jai Press Inc

Australian Vice-Chancellors' Committee (AVCC). The progress of higher degree students. Canberra: Australian Vice-Chancellors' Committee, July 1990a

Australian Vice-Chancellors' Committee (AVCC). Code of practice for maintaining and monitoring academic quality and standards in higher degrees. Canberra: Australian Vice-Chancellors' Committee, 1990b

Ballard B (1996) Contexts of Judgement: An analysis of some assumptions identified in examiners' reports on 62 successful PhD theses. Paper presented at the quality in postgraduate research conference, Adelaide

Baron R, Misovich S (1999) On the relationship between social and cognitive modes of organization. In: Chaiken S, Trope Y (eds) Dual-process theories in social psychology. New York: Guilford, pp586–605

Barrett E, Magin D, Smith E (1983) Survey of postgraduate research students enrolled at the University of New South Wales. Unpublished Research Report. Sydney: Tertiary Education Research Centre, University of New South Wales

Bazerman C (1981) What written knowledge does. Philosophy of the Social Sciences 2:361–87

Bazerman C (1987) Codifying the social scientific style: The APA publication manual as a behaviourist rhetoric. In: Nelson JS, Megill A, McCloskey DN (eds) The rhetoric of the human sciences. Madison, Wisconsin: University of Wisconsin Press, pp125–44

Bazerman C (1988) Shaping written knowledge. Madison, Wisconsin: University of Wisconsin Press

Bazerman C (1995) Influencing and being influenced. Social Epistemology 9(2):189–201

Becher T (1981) Towards a definition of disciplinary cultures. Studies in Higher Education 6:109–22

Becher T (1984) The cultural view. In: Clark B (ed) Perspectives on higher education: Eight disciplinary and comparative views. Berkeley: University of California Press, pp165–98

Becher T (1987a) The disciplinary shaping of the profession. In: Clark B (ed) Academic profession: National, disciplinary and institutional settings, ch. 6. Berkeley: University of California Press

Becher T (1987b) Disciplinary discourse. Studies in Higher Education 12:261–74

Becher T (1989a) Academic tribes and territories: Intellectual enquiry and the cultures of the disciplines. Milton Keynes: Open University Press

Becher T (1989b) Historians on history. Studies in Higher Education 14:263–78

Becher T (1990a) Physicists on physics. Studies in Higher Education 15:3–19

Becher T (1990b) The counter-culture of specialisation. European Journal of Education 25:333–46

Becher T (1993) Graduate education in Britain: The view from the ground. In: Clark BR (ed) The research foundations of graduate education, ch. 4. Los Angeles: University of California Press

Becher T, Henkel M, Kogan M (1994) Graduate education in Britain. London: Jessica Kingsley

Becher T, Trowler PL (2000) Academic tribes and territories: Intellectual enquiry and the culture of disciplines (2nd edn) Buckingham, UK: SRHE

Becher T, Parry S (2005) The endurance of the disciplines. In: Bleikle I, Henkel M (eds) Governing knowledge, ch. 9. Dordrecht, The Netherlands: Springer

Becker H, Carper J (1956) The development of identification with an occupation. The American Journal of Sociology 61:289–98

Berg RA (1992) Equity and non equity cooperative agreements: Implications for small business performance. Unpublished PhD thesis. University of Auckland

Biggs J (1999) Teaching for quality learning at university. Buckingham: SRHE and Open University Press

Biglan A (1973a) The characteristics of subject matter in different scientific areas. Journal of Applied Psychology 57:195–203

Biglan A (1973b) Relationships between subject matter characteristics and the structure and output of university departments. Journal of Applied Psychology 57:204–13

Blackburn RT, Chapman DW, Cameron SM (1981) 'Cloning' in academe. Research in Higher Education 15:315–27

Bourner T, Bowden R, Laing S (2001) Professional doctorates in England. Studies in Higher Education 26(1):65–83

Bourdieu P (1977) Outline of a theory of practice. Cambridge: Cambridge University Press

Bourdieu P (1988) Homo academicus. Translated by Peter Collier. Stanford University Press

Bourdieu P (1990) The logic of practice. Cambridge: Polity Press

Bowen WG, Rudenstine NL (1992)_In pursuit of the PhD. Princeton, New Jersey: Princeton University Press

Braxten JM, Hargens LL (1996) Variation among academic disciplines: Analytical framework for research. Higher education: Handbook of theory and research. New York, Agathon Press

Brennan J, Henkel M (1988) Economics. In: Boys CJ, Brennan J, Henkel M, Kirkland J, Kogan M, Youll PJ (eds) Higher education and the preparation for work. London: Jessica Kingsley, pp93–110

Brennan J, Shah T (2000) Quality assessment and institutional change: Experiences from 14 countries. Higher Education 40(3):331–349

Brine J (2000) Defining an appropriate range of learning opportunities for an outward looking PhD. Paper presented at the professional doctorates 3rd Biennial international conference, Doctoral education and professional practice: The next generation, University of New England, Armidale, NSW Australia, pp10–12 September, 2000

Buckley PJ, Hooley GJ (1988) The Non-completion of doctoral research in management. Educational Research 30:110–20

Burgess RG (1994) Some issues in postgraduate education and training in the social sciences: An introduction. In: Burgess RG (ed) Postgraduate education and training in the social sciences.London: Jessica Kingsley Publishers, pp1–12

Burgess RG, Pole CJ, Hockey J (1994) Strategies for managing and supervising the social science PhD. In: Burgess RG (ed) Postgraduate education and training in the social sciences. London: Jessica Kingsley Publishers, pp13–33

Cartwright N, Cat J, Fleck L, Uebel T (1996) Otto Newrath: Philosophy between science and politics. Cambridge: Cambridge University Press

Chaiken S, Trope Y (eds) (1999) Dual-process theories in social psychology. New York: Guilford

Clark BR (1963) Faculty culture. In: Lunsford TF (ed) The study of campus cultures. Boulder, Col: Western Interstate Commission for Higher Education

Clark BR (1992) The distinctive college. New Brunswick, New Jersey: Transaction Publishers. (First published by Aldine Publishing Company, 1970)

Clark BR (1980) Academic culture. Yale higher education research group working paper No. 42, March, 1980

Clark BR (1983) The Higher Education System. Berkeley, California: University of California Press

Clark BR (1987b) The academic life: Small worlds, Different worlds. Princeton, New Jersey: The Carnegie Foundation for the Advancement of Teaching

Clark BR (ed) (1987c) The academic profession national, Disciplinary and institutional settings. California: University of California Press

Clark BR (ed) (1993) The research foundations of graduate education. Los Angeles: University of California Press

Connell RW (1985) How to supervise a PhD. Vestes 2:38–41

Crane D (1972) Invisible colleges. Chicago: University of Chicago Press

Cronin B (1984) The citation process. London: Taylor Graham

Cullen D, Pearson M, Saha L, Spear R (1994) Establishing effective supervision. Canberra: Australian Government Publishing Service

Delamont S, Eggleston JF (1983) A necessary isolation? In: Eggleston J, Delamont S (eds) Supervision of students for research degrees. Proceedings of the British educational research association conference

Delamont S, Atkinson P, Parry O (2000) The doctoral experience: Success and failure in graduate school. London: Falmer Press

Delanty G (2001) Challenging knowledge: The university in the knowledge society. Buckingham: Open University Press

Dill D (1982) The management of academic culture: Notes on the management of meaning and social integration. *Higher Education* 11:303–20

Dunn AM (1993) The song of the Zebra Finch: Contexts and possible functions. Unpublished PhD. La Trobe University

Elton L, Pope M (1989) Research supervision: The value of collegiality. Cambridge Journal of Education 19:267–75

Epstein S, Pacini R (1999) Some basic issues regarding dual-process theories from the perspective of cognitive-experiential self-theory. In: Chaiken S, Trope Y (eds) Dual-Process theories in social psychology. New York: The Guilford Press, pp462–482

Fleck L (1979) Genesis and development of a scientific fact. Chicago: University of Chicago Press

Fulton O (1996) Which academic profession are you in? In: Cuthbert R (ed) Working in higher education. Buckingham, UK: Society for Research into Higher Education & Open University Press, pp157–169

Geertz C (1973) The interpretation of cultures. New York: Basic Books

Geertz C (1983) Local knowledge. New York: Basic Books

Gellert C (1993a) The German model of research and advanced education. In: Clark BR (ed) The research foundations of graduate education. Berkeley, California: University of California Press, pp5–44

Gellert C (1993b) The conditions of research training in contemporary German Universities. In: Clark BR (ed) The research foundations of graduate education. Berkeley, California: University of California Press, pp45–68

Gerholm T (1990) On tacit knowledge in academia. European Journal of Education 25:263–71

Gibbons M, Limoges C, Nowotny H, Schwartzman S, Scott P, Trow M (1994) The new production of knowledge, London: Sage

Gilbert G (1977) Referencing as a form of persuasion. Social Studies of Science 7:113–22

Gilbert GN, Mulkay M (1984) Opening pandora's box. Cambridge: Cambridge University Press

Gillett S (1990) Contemporary departures from realism: Placing Australia. Unpublished PhD thesis. La Trobe University

Gould AJ(1989) The Melbourne symphony orchestra: Factors influencing its character and identity. Unpublished PhD thesis. La Trobe University

Gumport PJ (1993a) Graduate education and organized research in the United States. In: Clark BR (ed) The research foundations of graduate education. Berkeley, California: University of California Press, pp225–60

Gumport PJ (1993b) Graduate education and research imperatives: Views from American campuses. In: Burton RC (ed) The research foundations of graduate education. Berkeley, California: University of California Press, pp261–96

Halliday MAK (1994) An introduction to functional grammar. 2nd edn. Melbourne: Edward Arnold

Hamilton DL, Sherman SJ, Maddox KB (1999) Dualities and continua implications for understanding perceptions of persons and groups. In: Chaiken S, Trope Y (ed) Dual process theories in social psychology. New York: Guildford Press, pp606–626

Harman G (2002) Producing PhD Graduates in Australia for the Knowledge Economy Higher Education Research and Development, 21(2):177–202

Harman KM (1988) The symbolic dimension of academic organization: Academic culture at the University of Melbourne. Unpublished PhD thesis. Melbourne: University of Melbourne

Harriott VJ (1983) Reproductive ecology and population dynamics in a scleractinian coral community. Unpublished PhD thesis. James Cook University

Hartnett RT, Katz J (1977) The education of graduate students. *Journal of Higher Education* XLVIII: 646–64

Hemmings BC (1994) Senior secondary school persistence and attrition: The development and testing of a theoretical model. Unpublished PhD thesis. University of New South Wales

Henkel M (2000) Academic identities and policy change in higher education. London: Jessica Kingsley

Henkel M, Kogan M (1993) Research training and graduate education: The British macro structure. In: Clark BR (ed) The research foundations of graduate education. Berkeley, California: University of California Press, pp71–114

Hockey J (1991) The social science PhD: A literature review. Studies in Higher Education 16:319–32

Kawashima T, Maruyama F (1993) The education of advanced students in Japan: Engineering, physics, economics, and history. In: Clark BR (ed) The research foundations of graduate education. Berkeley, California: University of California Press, pp326–354

King AR, Brownell J (1966) The curriculum and the disciplines of knowledge. New York: Wiley

Knorr KD (1977) Producing and reproducing knowledge: Descriptive or Constructive? Social Science Information 16:669–96

Knorr-Cetina KD (1981a) The manufacture of knowledge. Oxford: Pergamon

Knorr-Cetina KD (1981b) Social and scientific method or what do we make of the distinction between the natural and the social sciences? Philosophy of Social Science 11:335–59

Knorr-Cetina KD, Mulkay M (eds) (1983) Science observed. London: Sage

Knorr-Centina, KD (1983) Science observed: Perspectives on the social study of science Beverly Hills, Sage Publications

Knorr-Cetina KD (1999) Epistemic cultures: How the sciences make knowledge. Boston: Harvard University Press

Kogan M (2001) Modes of knowledge and patterns of power. Paper presented at the UNESCO Forum on higher education research, Research on knowledge. Paris

Kolb DA (1981) Learning styles and disciplinary differences. In: Chickering A (ed) The modern American college. San Francisco: Jossey Bass pp232–55

Kolb DA (1984) Experiential learning. Engelwood Cliffs, New Jersey: Prentice-Hall

Kuhn TS (1970) The structure of scientific revolutions, 2nd edn. Chicago: University of Chicago Press. (First published 1962)

Lammers C (1974) Mono- and Poly-paradigmatic developments in natural and social sciences. In: Whitley R (ed) Social processes of scientific development. London: Routledge and Kegan Paul, pp123–47

Latour B, Woolgar S (1979) Laboratory life. The social construction of scientific facts. Beverly Hills: Sage Publications

Latour B (1987) Science in action. Cambridge, Massachusetts: Harvard University Press

Lave J, Wenger E (1991) Situated learning: Legitimate peripheral participation. Cambridge: Cambridge University Press

Lerner MJ (1987) Integrating societal and psychological rules of entitlement: The basic task of each social actor and fundamental problem for the social sciences. Social Justice Research 1:107–125

Leslie SL (1993) 'A friendly debate': Religious and political discourse in restoration England, 1668–1674. Unpublished PhD thesis. La Trobe University

Levy SR, Plaks JE, Dwek CS (1999) Modes of social thought: Implicit and social understanding. In: Chaiken S, Trope Y (eds) Dual process theories in social psychology. New York: Guildford Press, pp179–202

Light DR (1974) Introduction: The structure of the academic profession. Sociology of Education 47:2–28

Linden J (1999) The contribution of narrative to the process of supervising PhD Students. Studies in Higher Education 24(3):351–369

Lodahl JB, Gordon G (1972) The Structure of scientific fields and the functioning of university graduate departments. American Sociological Review 37:57–72

Long M, Hayden M (2001) Paying their way. Canberra: Australian Vice-Chancellors' Committee

McConchie D (1984) The geology and geochemistry of the Joffre and Whaleback Shale members of the Brockman iron formation. Unpublished PhD thesis. University of Western Australia

McWilliam E, Taylor P, Singh P (2002) Doctoral education, Danger and risk management. Higher Education Research and Development. Special edition on doctoral education in the knowledge economy. (2):119–129

McWilliam E, Taylor P, Green B, Maxwell T, Wildy H, Simons D (2002) Research training in doctoral programs what can be learned from professional doctorates? Evaluations and investigations programme, Commonwealth department of education science and training, Commonwealth of Australia

Mendelsohn E, Weingart P, Whitley R (eds) (1977) The social production of scientific knowledge. Dordrecht, Holland: D. Reidel Publishing Company

Merton RK (1957) Priorities in scientific discovery: A chapter in the sociology of science. American Sociological Review 22:635–59

Merton RK (1973) The sociology of science. Chicago: The University of Chicago Press

Miller CML, Parlett M (1976) Cue consciousness. In: Hammersley M, Woods P (eds) The process of schooling. London: Routledge and Kegan Paul, pp143–50

Mitroff II (1983) The subjective side of science. Seaside, California: Intersystems Publications. (First Published Amsterdam: Elsevier, 1974)

Moritz CC (1984) Comparative studies of evolution in *Gehyra* and *Heteronotia* (Gekkonidae: Reptilia). Unpublished PhD thesis. Australian National University

Morton RJ (1990) The enemy within the gates: the internment of Australian citizens during the great war. Unpublished PhD thesis. La Trobe University

Moses I (1984) Supervision of higher degree students – Problem areas and possible solutions. Higher Education Research and Development 3:153–65

Mullins G, Kiley M (2002) 'It's a PhD, Not a nobel prize': How experienced examiners assess research theses. Studies in Higher Education 27(4):369–386

Moses I (1992) Good supervisory practice. In: Moses I (ed) Proceedings from the ARC and AVCC sponsored conference on research training and supervision. Canberra: Australian Vice-Chancellors' Committee, pp11–15

Neave G (1993) Separation de Corps: The training of advanced students and the organisation of research in France. In: Clark BR (ed) The research foundations of graduate education. Berkeley, California: University of California Press, pp159–91

Neave G, Edelstein R (1993) The research training system in France: A microstudy of three academic disciplines. In: Clark BR (ed) The research foundations of graduate education. Berkeley, California: University of California Press, pp192–222

Ruth N (2002) Diversity, Doctoral education and policy. Higher Education Research and Development, Special Edition on doctoral education in the knowledge economy 21(2):167–178

Noble K (1994) Changing doctoral degrees: An internatiopnal perspective. Bristol: Taylor and Francis Publishers

Nowotny H, Scott P, Gibbons M (2001) Re-Thinking science knowledge and the public in an age of uncertainty. Cambridge: Polity Press

Pantin CFA (1968) The relations between the sciences. Cambridge: Cambridge University Press

Parry O, Atkinson P, Delamont S (1994) Disciplinary identities and doctoral work. In: Burgess RG (ed) Postgraduate education and training in the social sciences. London:Jessica Kingsley, pp34–52

Parry S (1997) Doctoral study in its disciplinary context. Unpublished PhD Thesis, La Trobe University, Melbourne

Parry S (1998) Disciplinary discourse in doctoral theses. Higher Education 36(3):273–299

Parry S, Dunn L (2000) Benchmarking as a meaning approach to learning in online settings. Studies in Continuing Education 22(2):219–243

Parry S, Hayden M (1994) Supervising higher degree research students: An investigation of practices across a range of academic departments. Canberra: Australian Government Publishing Service

Parry S, Hayden M (1999)Experiences of supervisors in facilitating the induction of research higher degree students to fields of education. In: Holbrook A, Johnston S (eds) Supervision of postgraduate research in education. Review of Australian research in education No. 5. Australian Association for Research in Education, pp35–53

Phillips E, Pugh DS (1994) How to get A PhD. 2nd edn. Buckingham: Open University Press (First published in 1987)

Pitkethly A, Prosser M (1995) Examiners' comments on the international context of PhD theses. In: McNaught CM, Beattie K (eds) Research into higher education: Dilemmas, directions and diversions. Melbourne: Higher Education Research and Development Society of Australasia, pp129–136

Polanyi M (1961)Knowing and being. Mind. 70:458–470

Polanyi M (1983) The tacit dimension. 2nd edn. Glouster, Massachusetts: Peter Smith Publisher. (First published by Doubleday and Company, 1966)

Powles M (1984) The role of Postgraduates in Australian Research. Report for the council of Australian postgraduate associations. Melbourne: Centre for the Study of Higher Education, University of Melbourne

Powles M (1988a) Know your PhD students and how to help them. Melbourne: Centre for the Study of Higher Education, University of Melbourne

Powles M (1988b) The problem of lengthy PhD candidature. Assistance for postgraduate students: Achieving better outcomes. Canberra: Australian Government Publishing Service, pp26–34

Powles M (1989a) How's the thesis going? Melbourne: Centre for the Study of Higher Education, University of Melbourne

Powles M (1989b) Higher degree completion and completion times. Higher Education Research and Development 8:91–101

Proudford CM (1992) The process of school improvement and the meaning of effective schooling: a contextual analysis of four high schools in New South Wales. Unpublished PhD thesis. University of New England

Reber AS (1993) Implicit learning and tacit knowledge: An essay on the cognitive unconscious. New York: Oxford University Press

Reber AS (1997) How to differentiate implicit and implicit modes of acquisition. In: Cohen JD, Schooler JW (eds) Scientific approaches to consciousness. New Jersey: Mahwah Press

Rip A (2000) Fashions, Lock-Ins and the heterogeneity of knowledge production. In: Jacob M, Hellstrom T (eds) The future of knowledge production in the academy. Buckingham: The Society for Research into Higher Education and Open University Press, pp28–39

Rothblatt S (1977) Newman's idea. History of Education Quarterly 17(3):327–344

Rudd E (1975) The highest education: A study of graduate education in Britain. London: Routledge and Kegan Paul

Ruscio KP (1987) Many sectors, many professions. In: Clark BR (ed) The academic profession. Berkeley: University of California Press, pp331–39

Simpson R (1983) How the PhD came to Britain: A century of struggle for postgraduate education. Guildford, Surrey: The Society for Research into Higher Education

Storer NW (1966) The social system of science. New York: Holt, Rinehart and Winston

Swales J (1983) Developing materials for writing scholarly introductions. In: Jordan RR, Collins W (eds) Case studies in ELT. London, pp188–200

Symes C (1999) Working for your future: the rise of the vocationalised university. Australian Journal of Education 43(3):241–253

Toulmin S (1972) Human understanding. Oxford: Clarendon Press

Trope Y, Gaunt R (1999) A dual-process model of overconfident attributional inferences. In: Chaiken S, Trope Y (eds) Dual process theories in social psychology. New York: Guildford Press, pp161–178

Tinkler P, Jackson C (2000) Examining the doctorate: Institutional policy and the PhD examinatiion process in the UK. Studies in Higher Education 25(2):167–180

Turner S (2000) What are disciplines? And how is disciplinarity different? In: Weingart P, Stehr N (eds) Practising interdisciplinarity. Toronto: University of Toronto Press, pp46–65

Usher R (2002) Diversity of doctorates in the knowledge economy. Higher Education Research and Development 21(2):143–153

Ushiogi M (1993) Graduate education and research organisation in Japan. In: Clark BR (ed) The research foundations of graduate education. Berkeley, California: University of California Press, pp299–321

Walker RF (1980) A semantic-syntactic analysis of child language within a functional frame. Unpublished PhD thesis. Griffith University

Wason PC (1974) Notes on the supervision of PhDs. Bulletin of the British Psychological Society 27:25–29

Wegener DT, Petty RE (1987) The flexible correction model: The role of naive theories of bias in bias Correction. In: Zanna MP (ed) Advances in experimental social psychology, vol 29. San Diego, California: Academic Press, pp141–208

Weingart P (2000) Interdisciplinarity: The paradoxical discourse. In: Weingart P, Stehr N (eds) Practising interdisciplinarity. Toronto: University of Toronto Press, pp25–45

Weinstock M (1971) Citation indexes, Encyclopaedia of library information science. New York: Marcel Dekker

Welsh JM (1979) The first year of postgraduate research study. Guildford, Surrey: The Society for Research into Higher Education

Welsh JM (1981) The PhD student at work. Studies in Higher Education 6:159–62

Welsh J (1982) Improving the supervision of postgraduate students. Research in Education 27:1–8

Wengar E (1998) Communities of practice. Cambridge: Cambridge University Press

Whitley R (1980) The context of scientific investigation. In: Knorr KD, Khron R, Whitley R (eds) The social processes of scientific investigation. London: D. Reidel Publishing Company, pp297–321

Whitley R (1984) The intellectual and social organisation of the sciences. Oxford: Clarendon Press

Woolgar S (1982) Discovery: Logic and sequence in a scientific text. In: Knorr KD, Khron R, Whitley R (eds) The social processes of scientific investigation. London: D. Reidel Publishing Company, pp239–68

Woolgar S (ed) (1982) Laboratory studies. In: Social Studies of Science 12:481–58

Youll P (1988) Physics. In: Boys CJ, Brennan J, Henkel M, Kirkland J, Kogan M, Youll PJ (eds) Higher education and the preparation for work. London: Jessica Kingsley, pp51–71

Youngman M (1994) Supervisors' and Students' experience of supervision. In: Burgess RG (ed) Postgraduate education and training in the social sciences. London: Jessica Kingsley Publishers, pp75–104

Zuckerman H, Merton RK (1971) Patterns of evaluation in science. Minerva 9:66–100

Zuckerman H, Merton RK (1973) Age, ageing and age structure in science In: Merton RK (ed) The sociology of science. Chicago: Il: The University of Chicago Press

INDEX

Academic conventions and traditions 17, 19–20, 22, 24–25, 32, 35–36, 42–45, 48, 50, 53, 59, 62, 66–68, 77–78, 81–83, 87–93, 98, 100–104, 107–111, 115–117, 122, 125–135, 141–150, 159–160, 162

Academic identity 26, 43, 107, 135, 163

Acknowledgements 22, 48, 88

Authorship 22, 73–76, 122–123, 145, 148

Benchmarking 99–102, 104–108, 134, 151, 155, 159, 171

Candidature, stages of 22, 42, 50, 91–92, 96

Citation, tacit rules of 22, 44, 77–78, 88, 111, 122–128, 145–149, 168, 173

Cognition 24, 30, 42–46, 48–49, 74, 84, 88, 94, 98, 104, 109, 112, 115, 127, 157

Collective research 23, 36, 43, 63, 73, 76, 82, 93, 107, 127, 134, 137, 139, 140, 148, 152

Completion rates 18, 30–31, 39, 42, 108, 139, 150–155, 159, 164

Completion times 18, 54, 108, 139, 150, 159, 172

Counter norms 22, 46–47, 49, 105, 149

Coursework 18, 21, 29–30, 40, 49, 57–58, 64, 137, 140, 155–158, 161

Coursework 18, 21, 29–30, 40, 49, 57–58, 64, 137, 140, 155–158, 161

Disciplinary culture 17, 20–21, 24–26, 35–41, 44–47, 66–67, 70, 81–82, 87, 104, 107, 133–136, 145, 158, 163–167

Disciplines 9, 22–27, 31–36, 39, 42–46, 49–50, 57–61, 65–677, 73, 77–81, 85–86, 89–90, 93, 100–105, 109–111, 119–146, 152–154, 160–172

Disciplines, cognitive aspects 141–142

Disciplines, dynamism and change 31–34, 96, 112, 146, 150

Disciplines, social aspects 46, 87

Individualistic fields of study 42–43, 54, 56, 59–63, 70, 78–80, 83, 99–101, 106–107, 110, 113–114, 117, 119, 136–141, 144–145, 148, 152, 159

Induction 7, 21–27, 26–27, 31, 46, 53–66, 104, 133–137, 141–142, 159–165, 171

Intellectual isolation 49, 69, 71, 138, 159

Intellectual rapport 137–139

Intellectual rapport 59, 137–139

Intellectual uncertainty 136, 148

Knowledge communities 32–35, 42, 67–68, 73, 104, 142, 146, 149, 158

Knowledge production 18, 21–23, 35, 40–43, 46, 49, 53–54, 59, 62–64, 67, 70–77, 90–98, 117, 133–135, 150–151, 163, 172

Knowledge production 18, 21, 23, 35, 40–43, 46, 49, 53–54, 59, 62–64, 67, 70, 73, 77, 90, 95, 98, 117, 133–135, 150–151, 163, 172

Learning, social 41, 46

Learning, tacit 22–23, 45–49, 68, 70, 82, 84, 86, 105–109, 129–130, 142–145, 149, 150

Metaphors 112–114, 120, 128

Norms 17, 21–22, 25–27, 31, 34–37, 42–49, 54, 61–65, 67, 69, 70, 76–77, 81–83, 86–87, 97, 102–107, 119, 126, 130, 140, 142–146, 149–153, 165–166

Organisational culture 22–23, 25, 40, 146–147

Peers, importance of 35, 47, 84–88, 95, 98, 101, 108, 110, 137, 142–145, 166

Savvy, cultural 48–49, 88, 90–92, 105, 122, 147, 165

HIGHER EDUCATION DYNAMICS

1. J. Enders and O. Fulton (eds.): *Higher Education in a Globalising World.* 2002
   ISBN Hb 1-4020-0863-5; Pb 1-4020-0864-3

2. A. Amaral, G.A. Jones and B. Karseth (eds.): *Governing Higher Education: National Perspectives on Institutional Governance.* 2002   ISBN 1-4020-1078-8

3. A. Amaral, V.L. Meek and I.M. Larsen (eds.): *The Higher Education Managerial Revolution?* 2003   ISBN Hb 1-4020-1575-5; Pb 1-4020-1586-0

4. C.W. Barrow, S. Didou-Aupetit and J. Mallea: *Globalisation, Trade Liberalisation, and Higher Education in North America.* 2003   ISBN 1-4020-1791-X

5. S. Schwarz and D.F. Westerheijden (eds.): *Accreditation and Evaluation in the European Higher Education Area.* 2004   ISBN 1-4020-2796-6

6. P. Teixeira, B. Jongbloed, D. Dill and A. Amaral (eds.): *Markets in Higher Education: Rhetoric or Reality?* 2004   ISBN 1-4020-2815-6

7. A Welch (ed.): *The Professoriate. Profile of a Profession.* 2005   ISBN 1-4020-3382-6

8. Å. Gornitzka, M. Kogan and A. Amaral (eds.): *Reform and Change in Higher Education. Implementation Policy Analysis.* 2005   ISBN 1-4020-3402-4

9. I. Bleiklie and M. Henkel (eds.): *Governing Knowledge.* A Study of Continuity and Change in Higher Education - A Festschrift in Honour of Maurice Kogan. 2005
   ISBN 1-4020-3489-X

10. N. Cloete, P. Maassen, R. Fehnel, T. Moja, T. Gibbon and H. Perold (eds.): *Transformation in Higher Educatin.* Global Pressures and Local Realities. 2005
    ISBN 1-4020-4005-9

11. M. Kogan, M. Henkel and S, Hanney: *Government and Research.* Thirty Years of Evolution. 2006   ISBN 1-4020-4444-5

12. V. Tomusk (ed.): *Creating the European Area of Higher Education.* Voices from the Periphery. 2006   ISBN 1-4020-4613-8

13. M. Kogan, M. Bauer, I. Bleiklie and M. Henkel (eds.): *Transforming Higher Education.* A Comparative Study. 2006   ISBN 1-4020-4656-1

14. P.N. Teixeira, D.B. Johnstone, M.J. Rosa and J.J. Vossensteijn (eds.): *Cost-sharing and Accessibility in Higher Education: A Fairer Deal?* 2006   ISBN 1-4020-4659-6

15. H. Schomburg and U. Teichler: *Higher Education and Graduate Employment in Europe.* Results from Graduates Surveys from Twelve Countries. 2006   ISBN 1-4020-5153-0

16. S. Parry: *Disciplines and Doctorates.* 2007   ISBN 1-4020-5311-8